SENSITIVE MATTER

SENSITIVE MATTER

FOAMS, GELS, LIQUID CRYSTALS, AND OTHER MIRACLES

Michel Mitov

Translated by Giselle Weiss

HARVARD UNIVERSITY PRESS
Cambridge, Massachusetts & London, England
2012

First published as *Matière sensible: Mousses, gels, cristaux liquides et autres miracles*, copyright © 2010 by Éditions du Seuil

Library of Congress Cataloging-in-Publication Data

Mitov, Michel.
    [Matière sensible. English.]
    Sensitive matter : foams, gels, liquid crystals, and other miracles / Michel Mitov ; translated by Giselle Weiss.
        p.   cm.
    Includes bibliographical references and index.
    ISBN 978-0-674-06456-0 (alk. paper)
        1. Soft condensed matter.   2. Colloids.   3. Rheology.   4. Materials.   I. Title.
    QC173.458.S62M5813 2010
    620.1'1–dc23   2011041643

CONTENTS

MATTER, ARE YOU THERE?

We began to suspect as much a century or two ago: Matter is no longer what it once was, that is, solid, concrete, insensitive, and practically indestructible—totally contrary to the mind.

Indeed, matter has changed whether we like it or not. It may be "dark" or "missing." It has not always been what we think it is. It has aged. It has a history, a personality; it harbors forces within itself. Dark matter is an immense vacuum, and its particles hold invisible terrors. Matter can be deceptive or malevolent—think "antimatter" or, in the context of dark energy, "quintessence." Which (incidentally) is not unlike the "spiritual matter" that preoccupied theologians during the Middle Ages.

Matter has lost its materiality. We have known so for a long time without realizing it. It is something we have learned to live with. Here is a book written by a "researcher in sensitive matter" (yes! it does exist) who tells us, simply and convincingly, that when we whip up a batch of mayonnaise, we are party to one of the deepest secrets in the universe. I now personally believe that the mystery of mayonnaise is at least as complex as the "miraculous" liquefaction of the blood of St. Januarius—a phenomenon that occurs in a church in Naples on certain occasions and only in the context of a particular ritual. In managing to mix oil and water (both contained in an egg), two resolutely irreconcilable materials, we achieve in our kitchen the miracle of mayonnaise without, however, organizing any procession or prayer in its honor. Perhaps that will come as a consequence of reading this book.

We also learn how American latex was transformed into rubber (though it hardly made the fortune of the obstinate Charles Goodyear), how paint dries on a canvas, why shampoo foams better the second time, and we touch on—but do not resolve—the prodigious enigmas of granular matter: sand, for example.

In the course of investigating sensitive matter—the goal of which is not its "transmutation" but rather an understanding of its amphilicity or, conversely, immiscibility (not as complex as its sounds: An amphiphile "loves both"; an immiscible material does not mix—anybody can identify with that). We will also explore our everyday world, which is less stable and predictable than it seems.

While reading these astonishing pages—and being repeatedly surprised that the paper did not just crumble under my fingers—I wondered, What if the same is true of the mind, which after all is a product of matter? What if some of our ideas are liquid and others soluble or foamlike? What if our deepest-seated convictions are in reality nothing but meso-morphic states, that is, transitory, intermediary? Could it be that the decisions we believe to be the most rational, most carefully developed, and most logical are in fact only provi-sional emulsions, a fog that quickly melts away? What if traditions were obese and philosophical thought a complex fluid? What if our faith were to melt at times? What if the future of knowledge already belongs to the nanoscience of the mind?

Perhaps, vis-à-vis our proudest possession—the suprem-acy of our mind over the rest of creation—we might soon be forced to accept that we function unawares like an oxy-moron, like a "liquid crystal," for example. Could it be that what defines life is a complex effervescence, a fugitive state, an instability, a solution of contradictions?

"Contain, stabilize, coat, thicken": These cooking terms are discussed here at such length—even Ferran Adrià, the renowned Spanish chef, comes in for a mention—that we would be forgiven for thinking that they also apply to the workings of our brain, whose incredibly active scullery—chemically driven, of course—remains a mystery to us. Moreover, assuming multiraciality is our future on Earth, then the matter of which we are made is eminently suited to it. It has even preceded us down the long road of impurity.

So many questions! Will we learn the finer points of syn-eresis? Will we take up research in thixotropy? Will we study zetetics? Will we end up as hybrid mixtures, or are we already there? Did Rabelais's *paroles gelées* portend reality? Do they hang suspended in the air, awaiting the intrepid sailors? Could we take them in our hands? Would we know how to decode them? Will we one day be able to clone the ineffable?

After thousands of years of humility, matter is now in a position to educate the mind. Or at least to teach it vocabulary. A fair return of words, if not of things.

Jean-Claude Carrière
Screenwriter *(Diary of a Chambermaid, Belle de Jour, That Obscure Object of Desire, The Milky Way, Taking Off, Cyrano de Bergerac, The Mahabharata)*

SENSITIVE MATTER, DIVINE MATTER?

The old Neapolitan woman went down on her knees to approach the body of the beheaded man and to piously collect his blood. She carefully filled two vials of it to take home. Then an old man drew near and placed the remains of the martyr in a trunk.

Time passed.

One day, as the trunk was being brought to Fuorigrotta, on the hills of Naples, to the catacomb at Capodimonte, it went through the Antignano area of the city, where the old woman lived. She shuffled toward the convoy and lifted the vials near the trunk. To the amazement of all, the dry, brown mass transformed into a perfectly fluid, dark-red liquid.

The martyr is *San Gennaro* (St. Januarius). He tops the list of fifty-odd patron saints of Neapolitans, who are grateful to him for having accorded them his protection against the plague and the ravages of Vesuvius. In 305 he fell victim to the great persecution carried out by the emperor Diocletian against the Christians. The old woman was a relative whose paralysis he had healed. She and her gesture of devotion were the beginning of the history of St. Januarius, enormously important to Naples, where the veneration of the saint's relics occurs three times each year in a ceremony during which the Neapolitans pray for the liquefaction of the blood: the Saturday preceding the first Sunday in May, the anniversary of the transfer of the martyr's remains; on September 19, the day of his death; and on December 16, the date Mount Vesuvius erupted in 1631, which *miraculously* spared the Neapolitans. A rapid liquefaction augurs well for Naples. However, if the phenomenon is delayed or does not happen at all, the city is beset with worry. Neapolitans are quick to cite the wretched years of nonliquefaction: 1527 (plague), 1973 (cholera), 1980 (an earthquake cracked several *palazzi* and caused massive panic).

Manipulating the dry residue of ordinary blood is not the way to liquefy it. The fluid dried out because its liquid phase evaporated, an irreversible transformation. The vial should at most have shown a dark brown solid (the clot) swimming in a rather colorless liquid (the serum) part of the plasma that remains fluid after coagulation. Moreover, pure blood would long ago have become corrupted and decomposed.

The material might be sensitive to other stimuli. In *Le Corricolo*[1]—a diary of his trip from Rome to Naples in 1835—Alexandre Dumas described at length the story of St. Januarius and the fervor that accompanies the Neapolitan *miracle*. In spring 1799 Naples had fallen to the French, and the new government wished to affirm its legitimacy—

The Miracle of St. Januarius, by Gabriele Castagnola, 1859.

"Green face! Yellow face! Won't you perform your miracle?" The traditional chant of anxious, ordinary Neapolitans, repeated in a mournful monotone, in the event of delayed liquefaction. The face of the reliquary bust is gold, hence "yellow." The epithet "green face" refers to tarnished gold.

what surer way than to organize a ceremony and let the saint decide? The general and his top advisors were carried with great pomp to the church of Santa Chiara. The blood was still perfectly coagulated at 8:30 in the evening, and the crowd grumbled and increased its hostility toward the occupiers. Let me pass the pen to Dumas to recount the scene where a priest is circulating near the adjutant:

> "A word, father, if you please," said the young officer.
>
> "Yes?" asked the priest.
>
> "I wish to say, on behalf of the general," ventured the adjutant, "that if the miracle does not occur within ten minutes, you will be shot in a quarter of an hour."
>
> The priest dropped the vial, which the young adjutant fortunately retrieved before it touched the ground, returning it with all due respect; then he rose and resumed his place beside the general.
>
> "Well?" said Championnet.
>
> "Well!" said the adjutant, "Have no fear, General. The miracle will happen within ten minutes."
>
> The adjutant spoke the truth, but he erred by five minutes. At the end of five minutes, the priest raised the vial and cried:
>
> "Il miracolo è fatto!"
>
> The blood was completely liquefied.

"What the hell?!" one might be tempted to exclaim but for the fear of blaspheming. Here is a material sensitive to its proximity to the body to which it once belonged, to the date of handling, and also—assuming we believe the novelist—to the injunction communicated to the handler. What remarkable behavior! An inviting subject for research and an experimentally rich one for both physicists and chemists. What atoms is this material made of, what molecules, what particles? What complex interactions take place under its sacred jurisdiction? We will go to Naples. Perhaps we, too, will be lucky enough to discover the most remarkable things where we least expect to find them.

SENSITIVE MATTER

LET US PRAISE SENSITIVITY, A UNIFYING VIRTUE

Why did an Egyptian scribe have only to add the merest dash of acacia sap to a precarious mixture of soot and water to create a remarkably stable ink? Why does a smidgen of detergent in a liter of water produce beautiful bubbles, whereas pure water does absolutely nothing? A tiny battery drives a liquid-crystal watch: Why does it require ridiculously little electrical energy to change the state of the screen every second for as long as two years? What is the key to making targeted, time-release drugs that reduce the risk of side effects? How does the cell membrane decide which substances to let through? Why would rubber be short lived were it not for a *soupçon* of sulfur? Can the liquefaction of St. Januarius's "blood" be reproduced in the laboratory? Why does a pile of sand look stable yet collapse at the slightest perturbation? Is it possible to make a foam that can be defoamed and refoamed at will? Why is the very fact of breathing a little-appreciated triumph repeated more than twenty thousand times a day? How can one egg yolk stabilize up to twenty-four liters of mayonnaise?

We could happily go on listing these materials. They all have something in common: A very slight perturbation, an apparently trivial modification to their chemical structure or composition disrupts one of their properties. The number of situations in which materials must adapt seamlessly to environmental conditions abounds! In the man-made world, sensitive matter is central to many industrial problems: emulsions, gels, soaps, plastics, liquid crystals, granular matter. In nature, life would not exist but for sensitive matter; all biological structures—red globules, proteins, membranes, to cite only a few of these "objects" from the living world—are based on this principle.

What is this state of matter whose basic character is ever ready to change? What will it change into? At whose behest? Especially, why do modesty and restraint govern the proceedings? Finally, what is the function of all this? What purpose does it have? Before traveling to London, Paris, Prague, Philadelphia, Moscow, Hautvillers, Barcelona, Saint-Maximin-la-Sainte-Baume, and Naples in search of the answers—by way of the Amazon, Ireland, Ancient Egypt, and Mauritania—led us head for the first of our destinations: the Balearic archipelago.

CONCILIATION

THE ART OF RESOLVING CONFLICTS

It is the depressingly familiar story of two parties at complete loggerheads. The source of the disagreement is a difference in point of view, of culture, of customs. One solution, quite simply, would be to part ways. However, a more constructive approach might be to mediate between the two entities to help them to live together and—why not?—to produce a shared creation. Utopia? Fool's imagining? Sensitive matter would counter with a resounding *No!* How can it be so sure?

CHAPTER 1

PEACEMAKING AMONG ENEMIES . . .

EASY WHEN A MEDIATOR IS INVOLVED

BLEND OR SEPARATE?

In 1756 the Duke of Richelieu's chef brought back from Port Mahon (a town in the Balearic Islands that the duke had conquered, in the military sense of the word) a recipe for a sauce based on olive oil and egg yolk. He called his discovery *mahonnaise;* later, it became *mayonnaise.* Of course, a story this specific about a preparation this famous is asking to be contested. Indeed, it is said, though less convincingly, that the word is derived from *magnonaise* (from *magner* or *manier*—to handle) or from *moyeunaise* (in the Middle Ages, the yolk of an egg was known as *moyeu).* It also (and still) appears that the inhabitants of la Mayenne and Bayonne make some claim to "parentage" based on phonetic proximity. At any rate, everybody would agree that oil does not mix with water; by this reasoning, it should not mix with egg yolk, either, which is 50 percent water. But why? Although, electrically speaking, a molecule of water is neutral, its atoms do carry charges: The single oxygen atom (O) has two negative charges, and each of the molecule's two hydrogen (H) atoms has a positive charge. Two water molecules unite when a hydrogen atom on one is attracted to the oxygen atom on the other, forming a hydrogen bond. Molecules that contain OH groups generally form hydrogen bonds with water molecules. Oil molecules, on the other hand, are triglycerides composed of carbon and hydrogen. Their structure resembles a comb with three teeth and has no space for OH groups, which returns us to the question of how to bring oil and water together.

CONCILIATION

Can it be done? Yes, because egg yolk[1] contains not only water but also lecithin molecules, which act as mediators. Each consists of two parts: one water loving (hydrophilic)

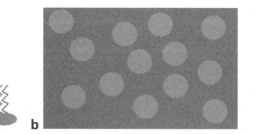

(a) Amphiphilic lecithin molecule: a head that likes water (but not oil) and two tails that like oil (but not water). (b) A successful mayonnaise: micelles of oil in water. The oil droplets are covered with lecithin molecules. These micelles are admitted by the water, in which they are well dispersed, because the lecithin molecules present their hydrophilic parts. (c) Cross-section of a micelle.

and the other (two-tailed) water hating (hydrophobic). We call these dual-affinity molecules amphiphiles from the Greek *philos* (friend) and *amphi* (both), which expresses the idea of a double possibility (an amphitheater has a left and a right side; an amphora has two handles). Amphiphilia will feature throughout our journey, whether the amphiphilic molecules be natural or synthetic and their functions biological or industrial.

In creating a stable mayonnaise with an oil concentration of more than 60 percent, lecithin molecules play a double role. They coat the drops of oil by linking their water-hating tails to them; the resulting "spheres"[2] are called micelles, from the Latin *mica,* for "morsel" (their thin-leaved structure enables mica fragments to be peeled apart). They also ensure the dispersion of the oil drops by exposing their hydrophilic heads to the water. The egg yolk proteins fulfill the same functions.

NO-FAIL MAYONNAISE
Micelles swell as oil is added, and the mix must be beaten all the while to break up the mass of oil into droplets. If too much oil is added, the drops coalesce into unequally sized larger drops. The mayonnaise then fails because there are no longer enough amphiphiles to protect all of the interfaces between the oil and the water. If there is too much yolk, the sauce will taste overwhelmingly of egg. The consistency will also be too firm owing to the closeness of the oil drops. Not enough water can ruin a mayonnaise, too, although adding a few drops of water, vinegar, or lemon juice while beating may resurrect it. Naturally, these measures also serve to soften the sauce, after which more oil can be added until it becomes too firm again, whereupon additional drops of

water can be added, and so forth. This is a practical way of gradually optimizing the quantity of amphiphilic molecules. If the concentration in the aqueous phase is too high, the result will be an emulsion of water in oil that risks catastrophic phase separation of the constituents.

Why not make a mayonnaise using egg whites? It is possible: They are 12 percent protein—the rest is water and mineral salts—and show good emulsifying properties. Beaten with oil, they form an emulsion . . . that is disgustingly inedible.

Micelles do not fuse: The hydrophilic heads of the amphiles are electrically charged, which causes them to repel each other. But the repulsion does not have to be strong: A little dose of salt or acid (vinegar, say, or lemon), incorporated at the beginning, is a good defense. The addition of these ions[3] weakens the charge of the amphiphiles, which in turn modulates the repulsion between micelles. When the liquid between the micelles is poor in ions, the repulsive forces are exerted directly, whereas when the solution contains many ions, the forces are weaker, which leads the micelles to coalesce. Here, too, the key is how much.

Mayonnaise made using an electric beater is less supple than that prepared with a fork because the micelles are smaller (rapid mixing divides the oil into smaller droplets), which increases their compactness at the microscopic level and thus the hardness of the emulsion. Some cooks recommend beginning the preparation when all the ingredients are at room temperature. When the oil or the egg yolk is cold, it is more viscous, which prevents proper formation of micelles. Similarly, egg whites just out of the refrigerator do not foam well because it is difficult to incorporate air into them. And use fresh eggs! As eggs age, their lecithin molecules degrade and lose the ability to make beautiful micelles. The same advice holds for emulsions of melted butter in an aqueous phase, such as béarnaise or hollandaise sauce (which differ by their seasoning and amount of butter).

A VERY DISCREET PEACEMAKER

Traditional recipes recommend using one to two deciliters of oil per egg yolk. Harold McGee, author of a respected encyclopedia on science in the kitchen, has prepared up to twenty-four liters of mayonnaise using a single egg yolk, injecting water as he added the oil because the amount of water in the egg yolk was insufficient to sustain the huge emulsion. And, remarkably, he did it without the help of extra lecithin, which shows very well that it takes a ridiculously small amount of amphiphiles to stabilize a mayonnaise and to progress spectacularly from immiscibility to miscibility. So if you find that egg yolk tastes unpleasantly of raw egg, you need use only a drop. The amount of amphiphilic molecules in the yolk is enough to cover a football field with a layer as thick as the length of the molecule.

BEYOND MAYONNAISE

Hervé This, a physical chemist and proponent of molecular gastronomy,[4] points out that genuine *aïoli* consists simply of crushed garlic to which olive oil has been progressively added. No egg yolk because the garlic contributes the amphiphilic molecules. Similarly, why not attempt an *oignoli* or an *échalotoli?* For onions and shallots also contain amphiphiles. It is possible to take culinary invention even further. In fact, all cells, whether plant or animal, are composed of amphiphiles (phospholipids). Hence *courgettoli* and *boeufoli,* which consist, in turn, of crushing a zucchini or a

On a London Pond

When speaking in public on the subject of amphiphiles (as during his acceptance speech to the Nobel Academy in 1991), physicist Pierre-Gilles de Gennes always told the story of how Benjamin Franklin measured the length of a molecule for the very first time. The success of the experiment was due both to the double affinity of amphiphiles and (indispensably) to their attraction at interfaces—in the case of mayonnaise, the water-oil interface and, in the case of Franklin's experiment, that of water and air.

In 1768, armed with a small spoonful of olive oil whose volume he had measured, Franklin found himself near a pond in London.[1] He tossed the oil into the water and could easily see how it spread: The surface of the pond was as smooth as a mirror. This experiment was part of Franklin's investigations into the power of oil to calm troubled waters (less well known than the inventor's studies of lightning and thunder),[2] something the Greeks had already discovered. After a while, the oil stopped spreading: It had reached a state of equilibrium. The main constituent of olive oil is an amphiphile known as oleic acid. In Franklin's experiment, part of the surface of the lake was covered with a single layer of amphiphilic molecules whose hydrophilic head was in contact with the water and whose hydrophobic tail was pointing away from the water. Franklin then estimated the amount of surface covered, and, dividing the volume of the spoon by the value of his estimate, he deduced that the layer of oleic acid was roughly three nanometers high, using today's units of measure. One nanometer—a millionth of a millimeter—is the ideal unit for describing the size of atoms and molecules; the size ratio between the planet Earth and an orange is the same as that between an orange and a nanometer-sized object. Although the value of three nanometers corresponds to the length of a molecule, Franklin did not report it as such, for he could hardly have grasped the significance of his observation. Indeed, the concept of molecule was not introduced until a more than a century later by physicist and Nobel laureate John W. Rayleigh.

The story is a lovely example of experiment based on simple reasoning, without sophisticated equipment, to obtain a rough estimate of the molecular length of an amphiphile. No need to contact a laboratory in Grenoble or Saclay[3] to set up a meeting to use synchrotron X-rays or a neutron beam from a reactor. Big instruments are wonderful—they are often indispensable—but, in evoking the spirit of Franklin, de Gennes hoped to encourage people to reflect on the diminished place of reasoning in an educational system that frequently favors formalism to the detriment of the sense of observation and critical thinking—a fortiori in the context of a science as experimental and deductive as physics. "I prefer ideas, even half-baked ones, to numbers, even precise ones,"[4] the physicist and philosopher Jean-Marc Lévy-Leblond has one of his characters remark on the behavior of matter.

Notes

1. Around Clapham Common.
2. B. Franklin, *Philosophical Transactions of the Royal Society,* 1774, vol. 64, pp. 445–460.
3. The locations of two major accelerator facilities.
4. J.-M. Lévy-Leblond, *Aux contraires,* Gallimard, 1996, p. 270.

cube of beef and then whisking it with oil. Any bit of vegetable or animal flesh can be turned into an emulsion sauce, although it might require adding a little water.

Having crossed the threshold of the kitchen with the aim of making things soluble, let us continue with dietary basics. Fats are insoluble in water, yet they are digested in water. How inconvenient! Mother Nature, however, deals with it by subterfuge.

CHAPTER 2

DISSOLVING FAT IN WATER:

A QUESTION OF ORGANIZATION

Bile is continually secreted by the liver and stored in the biliary vesicle, a little reservoir situated between the liver and the intestine, until the organism requires it for digestion. It is 90 percent water. The rest is a mix of bile salts, cholesterol, and lecithin (again!). In the 1960s, biophysicists Donald M. Small and Dikran Dervichian of the Pasteur Institute, in Paris, studied the behavior of bile as a function of its composition.[1] This led them to draw a triangular phase diagram as a practical way of representing the phases of a ternary (three-part) mixture for all the concentrations of the trio. Determining the corresponding concentrations for each point inside the triangle simply required drawing three lines parallel to the three sides and reading the values at the intersection of the axes. Small and

Dervichian showed that only mixtures whose composition was in the small area at the lower left of the triangle are homogeneous and capable of maintaining cholesterol in solution.

For example, the blend indicated in the figure on the next page—64 percent bile salts, 29 percent lecithin, and 7 percent cholesterol—corresponds to a phase in which the cholesterol remains well dispersed, like oil in mayonnaise. When the trio of concentrations is correct, the bile salts and the lecithin transform the cholesterol into micelles in the bile water because they are amphiphiles. Bile is a genuinely sensitive material: A slight increase in cholesterol or a decrease in salts may destroy the homogeneity of the mix. The cholesterol precipitates more in the vesicle than in the liver because bile is more concentrated there. Gallstones can vary in size from a grain of sand to a golf ball. They are caused either by a deficiency of bile salts or an excess of fat. The imbalance in turn may be the result of diet—for example, the "big eater" syndrome more typical of the United States and Europe than the Far East—or hormones: Gallstones are two to three times more frequent in women (estrogen increases the amount of cholesterol in bile) and appear to be related to number of pregnancies. Treatments have been developed that dissolve gallstones with bile acid, forming the precious salts in the presence of cations.[2]

Once fat and cholesterol have been transformed into an emulsion by bile salts in the intestines, enzymes can apply their efficiency to reducing the micelles to smaller bits, which promotes their absorption. The work of the enzymes is made easier by the division of clusters of big molecules of fat into millions of fine droplets. Emulsification has a two-fold function: dispersing the fat phase in water and exposing a larger portion of the fat's surface area to the action of the digestive enzymes.

Small and Dervichian were among the first to link the formation of micelles to human physiology—a mechanism as basic as digestion brought to light only forty years ago. Dervichian died in 1988. Small currently heads the Department of Physiology and Biophysics at Boston University. He remembers his advisor and friend well and with great affection as having been an outstanding scientist. Small also spe-

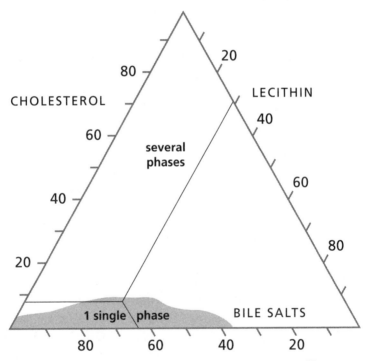

Small and Dervichian's triangle takes into account the homogeneous (the zone at the lower left) and inhomogeneous aspects of bile based on the concentration of each of its three components: bile salts, lecithin, and cholesterol.

cifically recalls an informal but crucial dinner conversation at a chemistry conference on surfaces held in Brussels in 1963, during which Alan Hoffman, a colleague and gastroenterologist, persuaded him to stick with developing his phase diagrams for elucidating the role of bile composition in the mechanism of digestion.[3] The anecdote is a testament to the vital role of informal contacts between scientists and a *magret de canard* washed down with burgundy in building knowledge.

Solubility by micelles is ubiquitous in life processes. It comes into play in industry, too, whenever a water-averse drug, perfume, or colorant needs to be dispersed in an aqueous medium. For example, with detergents, fatty residues are pulled from the surface and held in solution in water by virtue of the arrangement of the amphiphilic molecules at their periphery. Moreover, as French stage actress Florence Blot famously chirped in a 1970s' TV commercial for a dishwashing liquid: *"A few drops is all you need!"* Nothing could be truer, physicochemically speaking. A few cubic millimeters can cover a molecular monolayer surface of several square meters.

Bringing in outsiders is not the only way to mediate between enemies. Simply creating a link between the two parties may lead them to find a dramatic compromise on their own. Like liquid crystals on display . . .

CHAPTER 3

DON'T MIX, ASSOCIATE!

THOSE WERE THE DAYS . . .

If you had been at the German university of Prague on March 14, 1888, and just happened to be peering over the shoulder of Austrian botanist Friedrich Reinitzer (1857–1927), you would have seen that he was writing to German physicist Otto Lehmann (1855–1922) at Aix-la-Chapelle:[1]

> Encouraged by Dr. Zepharowich (professor of mineralogy at Vienna), I venture to ask you to investigate somewhat closer . . . / . . . the two enclosed substances. Both substances show such striking and beautiful phenomena that I can hopefully expect that they will also interest you to a high degree . . . The [first] substance has two melting points, if it can be expressed in such a manner. At 145.5°C it melts to a cloudy, but fully liquid melt which at 178.5°C suddenly becomes completely clear . . . / . . . The cloudiness on cooling is caused not by crystals but by a liquid which forms *oily streaks* in the melt.

Reinitzer proceeded to report the blue and violet color changes visible to the naked eye that he observed in chilling the substance to the point of crystallization. He described the behavior of cholesteryl benzoate crystals extracted from carrots and from gallstones. Now, suppose you were asked during a quiz: "What do a carrot and a flatscreen have in common?" Confident of the connection you have drawn between a multimillion-dollar industry and the whimsical observations of a botanist of yore, you would respond: "Liquid crystals, of course!" Scientists before Reinitzer had made similar observations in natural substances, for example, Rudolf Virchow (1821–1902) in discovering myelin[2] in 1854. However, their research had gone no further than hypotheses, probably because of the complexity of the biological structures they were studying. Reinitzer is credited with the discovery because he was the first to note *two melting points,* an inconvenient expression vis-à-vis the nomenclature of the time and one that would cause Reinitzer a fair amount of embarrassing confusion among his contemporaries: "What? A crystal that has two melting points? Meaning it melts at two separate stages?" The fact is that a crystal transforms into a perfectly clear liquid at a single, very well-defined temperature. Reinitzer's early detractors perceived artifacts in his observations: crystalline impurities in a colorless liquid or a two-phase emulsion. In fact, the contrarians ignored the characteristics described by the botanist—which were infinitely more interesting—and did not even try to prove their claims. Lehmann immediately extended Reinitzer's observations to a large number of natural substances and recruited his physical chemist colleagues to persuade the scientific community (by protracted debate) that what they were seeing was a homogeneous phase, a true stable state of matter between the crystalline and the liquid phases. The combination of optical properties typical of crystals and the liquid state led a hesitant Lehmann to name the material *flowing crystals, crystalline liquids,* and finally *liquid crystals,* an oxymoron from the past that has persisted to today's technologies. The unusual term *liquid crystals* thus stems from the confusion of these nineteenth-century scientists. Liquid crystals are the basis of light, ultraresistant[3] materials such as Kevlar, which have nothing *liquid* about them.

The French crystallographer Georges Friedel (1865–1933) tried to establish some order in the terminology. In 1922 he wrote a two-hundred-page[4] article for *Annales de Physique* on "mesomorphic states of matter," from the Greek *mesos* (intermediate) and *morphê* (form). Accordingly, he recommended applying this locution to the naming of states intermediate between crystal and liquid. In this epic saga—a veritable reference work—Friedel vehemently opposed the term *liquid crystals* as a source of confusion in understanding the nature of these most unusual materials, which "Lehmann had the great merit of drawing attention to . . . , but [which] he erred greatly in naming." Inadvertently, Lehmann had seriously diminished the importance of what Reinitzer and he had discovered. Friedel minced no words, regretting profoundly that Lehmann had called his materials " 'liquid crystals.' This term was soon accepted just about everywhere either by scientists who had not had the opportunity to see for themselves, or by those who had only vague ideas about crystals and crystallized matter." It would be hard to find such language in today's scientific literature, and any perpetrator of an *ad hominem* attack would be taken to task by the editor of the journal. *O tempora! O mores!*

Exactly what mesomorphic states of matter did Friedel mean? The essential phase, which we encounter the most often because it is used in flatscreens—from a simple watch, to the display of a computer screen, to the ubiquitous cellular telephone—is the *nematic* phase. Looked at under a microscope, a fine layer of this liquid crystal between two glass sheets has a threadlike structure. (The Greek word for thread is *nêmatos*.) In fact, Friedel and his coworker, François Grandjean (1882–1975), called these liquid crystals *thread liquids* and began to study them together in 1909. As early as the following year, they dared propose that "Lehmann liquids represent a new state of matter, as different from the crystalline state as from the state of ordinary liquid."[5]

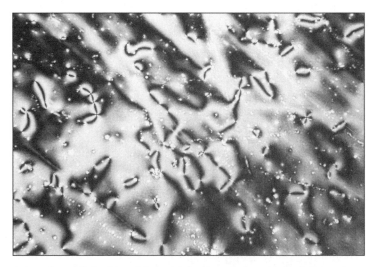

A drop of nematic liquid crystal, sandwiched between two sheets of glass and observed under the polarized light of an optical microscope, spontaneously presents a threadlike texture.

Grandjean—whose prodigious output included mineralogy, paleontology, and geology—was piling up observations. In 1918, while carrying out a microscopic investigation of the structure and defects of a nematic liquid crystal placed between two freshly cleaved crystalline sheets,[6] he accidentally discovered a key property: the effect of an electric field on liquid crystals.[7] He suspected that the phenomena he observed were caused by electrical charges on the surface of the sheets. Accordingly, he performed similar experiments using the electrodes of a generator to test his hypothesis. In 1913 Charles Mauguin (1878–1958) had described an orienting effect on liquid crystal molecules caused by contact with the same surfaces.[8] In short, from 1910 onward, these physicists had discovered the two fundamental effects at play in the liquid crystal displays that finally appeared in the 1970s. Patents were obtained much later—and elsewhere. The sudden interruption of Grandjean's work at the beginning of the 1930s says a lot about contemporary scientific values: "Because I could not procure sufficient funds prior to the last war to equip my laboratory at the École des Mines, I had to abandon my research on what we used to call liquid crystals,"[9] he wrote biologist Yves Bouligand in 1967 with some nostalgia.

The political decision makers of French research at the time, as well as a few self-congratulatory influential thinkers—some of whom preferred to see the subsidies flow into their own disciplines—considered liquid crystals to be a "laboratory curiosity" that would never find any application. The researchers were encouraged to turn their attention to more fashionable subjects such as quantum mechanics and nuclear physics if they wished to have any hope of financial support. How risky and presumptuous to declare the ideas of a discipline to be so clear-cut before the fact, particularly when people had been working in a field for only a short period of time! Significant research, albeit rare and isolated, continued in Europe. However, liquid crystals had already begun to cross the desert, a migration that would last until the end of the 1950s. History has shown that the lack of interest was undeserved. Doubly so for liquid crystals because what was known about them did not even appear in textbooks on the structure and properties of materials; consequently, future scientists and teachers of the period had no idea they even existed. The history of research in the area of liquid crystals—though other areas would suffer similar vicissitudes—is worth reflecting on in an era where many teams are rushing headlong into *nanoscience*.[10] Nonetheless, there are still many problems whose solution depends on mechanisms that occur at other levels.[11]

Some years ago, at public conferences on liquid crystals, I used to say that it was pioneering researchers (Friedel, Mauguin, and Grandjean) who had christened the nematic state—until I learned the real history, which arose from much less formal circumstances. Physicist Jacques Friedel is an expert in the structures and defects of metals and metal alloys, and, together with André Guinier, he founded a research school in solid-state physics. Friedel was very familiar with the history of liquid crystals, which he described in his memoirs.[12] One afternoon, when Jacques's grandfather Georges Friedel was relaxing with his daughters, he showed them photographs of liquid crystals taken under a microscope. One of them (Marie Friedel) shouted with delight on seeing the abundance of threadlike structures, "Father! The Greek word for thread is *nêmatos*. You should call your liquid crystal 'nematic'!"[13] Marie, who never graduated from high school, went on to teach her nephew Jacques Greek and hieroglyphics when he was barely ten years old. She devoted her life to her family, helped her father, and maintained her passion for ancient languages to her dying day. This story is little known among researchers in the field.

One day when I was attending the annual international conference on liquid crystals in Edinburgh, during a coffee-break conversation I asked an eminent professor the origin of the term. He referred to Georges Friedel and his observations about threadlike textures. I answered mischievously though respectfully: "No! Marie Friedel— during an afternoon outing with her father." To my sur-

Georges Friedel's (left) amateur string quartet at the École des Mines de Saint-Étienne. The little girl at left rear is Marie, "a pale and diminutive figure" (J. Friedel, correspondence with the author, January 5, 2009).

prise, my colleague's face clouded over. My alternative account, which I find charming, seems to have bordered on the sacrilegious.

Friedel's contribution was hardly limited to terminology. He demonstrated the molecular organization of the principal phases of liquid crystals through microscopic examination of their texture and defects as revealed by the fine layers of these *mesomorphic states of matter.* His predictions were later confirmed by X-ray studies carried out by his son, Edmond.

So much for the work of researchers on liquid crystals during a pivotal period at the turn of the century. What did the next generations make of them?

THE PSYCHOLOGY OF A SCHOOL OF FISH

Between order and disorder reigns a delicious moment.

—Paul Valéry[14]

A nematic liquid crystal is not composed of just any molecules. They must be elongated. If for only this reason, it is impossible to produce a liquid crystal phase by heating or chilling water. How are the molecules arranged with respect to each other? Like a school of fish. These fish molecules have a common orientation, yet no position. *Elongated molecules presenting a disordered position but an orientational order:* There's an advertisement for a liquid crystal to display on a flatscreen.

The optical properties of a nematic liquid crystal, which are what relates it to crystal, derive from its orientational order. If this order did not exist, it would be impossible to exploit nematic liquid crystals for display technologies. The property of fluidity—associated with liquid—enables the orientation of the molecules to change by the application of a slight voltage. The electrical power required to drive a basic liquid crystal screen—one that displays characters and numbers—is on the order of a millionth of a watt (the bulb in a bedside lamp consumes a few tens of watts). The battery has to be replaced only every two or three years. The portability of flat liquid crystal screens, as opposed to plasma screens, is the result of this modest appetite for energy.

Analogous to a school of fish—why not? But fish are always moving! Nematic molecules move, too, owing to perpetual thermal agitation.[15] Fish constantly readjust their relative orientation, and so do the molecules. But the analogy goes further. In the image shown here, a snapshot of a

A nematic school of fish: the fishlike molecules are all oriented more or less in the same direction, whereas their position is disordered.

moving school, at the lower right, a little group of contestants has adopted an orientation different from the rest of the school. In fact, as a result of dust or scratches on the surface on which the liquid crystal rests, the orientation of the molecules is not homogeneous: Defects appear, and each thread of the texture corresponds precisely to the location of a defect. A thread marks the site of the interface between two little groupings of fish in slight disagreement over which orientation to adopt. For a screen to function perfectly, the orientational order of the group must be perfect, which the manufacturer tests by covering the glass plate with a very thin sheet of plastic film that is then brushed in a single direction that one wishes to impose on the nematic molecules. The solution to the problem required much research before the commercialization of the first watch screens. *Interactions between surfaces and liquid crystals* continue to fuel a great deal of research, all the more complex and exciting when the surface is nonplanar or flexible. This last possibility is particularly interesting because it is key to the flat, flexible screens of the future: Magazines we will be able to fold and unfold at will, while downloading the newspaper of our choice.

MOLECULAR ANATOMY

A molecule of heptane (a common laboratory hydrocarbon and good for dissolving fat) is very flexible. Its structure is

Two typical components of the fishlike liquid crystal molecule of our flatscreens. Left: a long flexible heptane chain. Right: a short, rigid cyanobiphenyl molecule.

simple: a chain of carbon and hydrogen atoms: CH_3-CH_2-CH_2-CH_2-CH_2-CH_2-CH_3. A cyanobiphenyl molecule also has a simple structure, but, in contrast, it is very rigid, consisting of a pair of benzene molecules hooked to a small *cyano* group made up of an atom of carbon and an atom of nitrogen. Benzene is a natural constituent of crude oil that is widely used in the chemical synthesis of drugs, dyes, and plastics. In a molecule of benzene—C_6H_6—the six carbon atoms form a ring.

In equal measure, heptane and cyanobiphenyl are miscible only above 77°C; the resulting phase is liquid. Below that temperature, they separate into one liquid phase rich in heptane and another crystalline phase of almost pure cyanobiphenyl. This demixing reflects the chemical incompatibility between the two substances. Their "disagreement" is total. One molecule is malleable and adaptive, whereas the other is rigid, jealous of its structural integrity, and electrically polarized in the bargain. Although, somewhat predictably, each substance produces both a crystal phase and a liquid phase when heated, their melting temperatures are very different: −91°C for heptane and 88°C for cyanobiphenyl.

How can they be made miscible? Create a chemical link between the two by attaching a chain of heptane to a

The conciliatory molecule: the flexible and rigid parts have been chemically linked, and agreement (partial but acceptable) is reached. A group of these molecules gives rise to a nematic liquid crystal state: the child of compromise.

cyanobiphenyl. Thereafter, everything changes. The compound is stable and becomes a liquid crystal. Precisely between 28.5 and 42°C it exhibits a nematic phase, and our desire to marry heptane and cyanobiphenyl is gratified.

Indeed, separation has become impossible: The two parties are bound—and for the better. Between 28.5 and 42°C, the flexible chains of heptane, weakly attracted to each other, create disorder, whereas the little cyanobiphenyls are strongly attracted and try to organize themselves, in other words, to create order. Whence the nature of compromise: a little order, fine—to be specific, orientational order, which will satisfy the cyanobiphenyls—but not too much, consequently positional disorder afforded by the flexible heptane chains. How better to define the nematic state? This rigid-flexible duality is the basis of the "step-by-step" melting between crystalline and liquid phases (the "two melting points" of Reinitzer's letter to Lehmann): The molecules lose their existing positional order in the crystalline phase at an initial temperature but maintain their orientational order, which they ultimately lose at a higher temperature. Here is how, owing to a chemical trifle—a "simple" bond—it is possible not just to morph from an immiscible system to a miscible one but also, more important, to create a new state of matter.

Makers of liquid crystal screens use a mix of ten or twenty different molecules because they must adjust several parameters simultaneously, and no single substance embodies all of the required values. The screen must function at the North Pole just as it does at the equator; accordingly, the transition temperatures of the liquid crystal are controlled by a mix of different molecules in which the length of the flexible chain and the atomic composition of the rigid parts vary. Manufacturers seek a nematic state across the widest possible range of temperatures and choose depending on the application—typically between 0 and 40°C for a household flatscreen TV. The molecule I described above was named 7CB by the company that synthesized it ("7" for the number of heptane hydrocarbons and "CB" for cyanobiphenyl); it comes into play in the composition of many nematic blends used in simple displays. In addition to the rodlike shape, there are also molecules shaped like disks and pyramids that give rise to superb liquid crystal phases. Except for its dual rigid-flexible character, in terms of chemical affinity a liquid crystal molecule often combines two opposites—hydrophilic and hydrophobic groups, as in an amphiphilic molecule—or a combination of groups (a hydrocarbon part attached to a fluorocarbon part).

A thin nematic film between two glass sheets presents a deceptive transparency to the naked eye, sort of like a film of water or of alcohol. However, the deception is a product of observation; liquid crystals have hidden properties that we will bring to light by encapsulating them in a certain way.

SEQUESTERING A SCHOOL OF FISH

Encapsulating liquid crystals consists of dispersing them in droplets several micrometers in diameter in a plastic sheet. To draw an analogy, it is like filling the holes of Gruyère cheese with liquid crystals. One has only to prepare a resin mix containing around 10 percent nematic liquid crystals, shake the mixture well, and deposit a thin film between two sheets of glass. At this moment, the film is still transparent. Then the resin is hardened either by heating it or by shining ultraviolet light on it (as at the dentist). Now the composite material is no longer transparent but opaque. In one droplet, the molecular fish are lined up parallel. However, because the diameters of the droplets are different, there is no reason for the average orientation of the molecules in one droplet to be the same as in another. This is why ambient light is scattered in all directions in space and why the material looks milky; similarly, clouds are white because light is scattered by ice particles or water drops of various sizes. It is possible to change the material from opaque to transparent by applying a voltage. Each of the cyano groups has an electric dipole—a pair of two electrical charges, one positive and one negative. When a voltage is applied—about ten volts, say—all the dipole molecules in all the droplets align in the same direction, that is, the direction of the electric field, and perpendicular to the surfaces of the glass. The effect is similar to the alignment of a compass needle with Earth's magnetic field. As the cause of the light scattering is removed, the material becomes transparent. If the voltage is cut, the liquid crystal relaxes because the molecules lose the

Left: confinement of a nematic school of fish in aquariums in the form of spherical cavities encapsulated in a plastic sleeve. The black lines represent the orientation of the molecules. There is no relation between the orientation of one sphere and that of another. Right: a voltage is applied. It has the effect of imposing a single and identical orientation on all of the molecules in every cell—here, perpendicular to the surfaces.

direction imposed on them. The imposer of order is no longer there, and the cell becomes opaque again.

WINDOWS WOULD DO WELL TO REFLECT

Going from an opaque to a transparent state and vice versa by applying, then removing a voltage is a property that has been put to good use in windows with variable levels of transparency. "An imaginary window!" cried a small girl observing a prototype that I was toggling at a conference. These windows are used for bathrooms, for security windows at banks, and as partitions for meeting rooms in offices.

The story of smart windows will perhaps not stop there. Other liquid crystals have the characteristic of reflecting one part of the incoming light and not another; in other words,

Smart glass. Left: in its native, milky, and opaque state. Right: transparent as the result of applied voltage

they are selective. This is not the case with variable-opacity windows, which modulate all of the incoming light by scattering it. One can imagine a window that lets in some of the sun's rays but not others, or that adjusts intensity. These two properties are driven by voltage. The user, or a little sensor placed on the outside, could choose the proper voltage based on brightness and temperature. In summer, one might find oneself in a room with a closed window and the right light but little heat because infrared rays have been blocked. This would save energy because air-conditioning would not be needed to regulate the temperature. Other scenarios are possible by altering the tint of the windows to attenuate visible light.

Jean Cocteau said, "Mirrors would do well to reflect further before throwing back images."[16] Some decades later, one is surprised to think, smiling, that a variably reflecting mirror sending back some of its rays to the sun depending on the circumstances, its humor, or its programming could be understood as a very technological translation of the poet's immaterial thought.

THE LITTLE ADDITIVE THAT CHANGES EVERYTHING

A lone individual sets a goal, begins to succeed, and then fails. Some missing element prevents this person from realizing his or her potential. How to solve this problem?

A group of people neither agree nor disagree. Nothing happens between them, that's all. They do not know that they could, however, do something great. How to clue them in on this possibility?

In both cases, an opportunity exists, but some little thing gets in the way of seizing it. It is not so much a conciliator that is called for here but rather an organizer or an orchestra conductor. It is up to sensitive matter to find this enabler.

CHAPTER 4

RUBBER:

A STORY NEARLY CUT SHORT

EXPLORING THE AMAZON

Imagine a scene taking place two thousand years ago, perhaps more, against the backdrop of an Amazonian forest in Brazil. Its inhabitants, the Amerindians, tap the trunk of a *Hevea* tree, collect the juice—whitish, a little viscous—and drench their feet with it. Neither sap nor resin—fig trees and dandelions secrete a similar substance—the liquid in question is called latex. Its molecules are very long, flexible chains known as *polymer,* from the Greek *polus* (many) and *meros* (part). Each molecule results from the repetition of a smaller constituent molecule, or monomer (*monos* [single]), which hooks up chemically with another identical monomer, which in turn links to another identical monomer, and so forth many, many times—hundreds or thousands of times.

Our latex is a little like a plate of tangled spaghetti-like molecules; if you bring your lips close to the end of a strand of spaghetti and suck in, you dissociate it from its neighbors. The viscosity of latex increases until the material solidifies within twenty minutes or so, thus shoeing the user with boots made to fit. This phenomenon results from the action of oxygen: It links the molecular chains one to the other, but only at a few points. To return to the example of the spaghetti, it is as though you sucked in the entire plate at once! The latex has changed from a viscous liquid state to an elastic solid one: It has become rubber.

However, on a smaller scale, that of molecules, the rubber is fluid. This duality—fluid at the microscopic level but macroscopically solid—gives rubber its special mechanical properties. Such rubber boots are a great invention, but they do not last long. After one day they fall apart. This rubber has no stability at all because oxygen acts on it in two contradictory ways. First, it helpfully creates bridges between the chains. Then it continues its activity but by cutting the chains. Apparently Christopher Columbus was the first European to take part in a game played by the Aztecs using a rubber ball. More formally, the material was rediscovered in 1735 by a French naturalist, Charles Marie de La Condamine, at a time when the Academy of Sciences was sending scholars to South America to test Newton's theories about the size and shape of the globe. As it turned out, La Condamine and his colleagues discovered quinine, curare, and rubber. They named it *caoutchouc* in French after hearing it pronounced *cao* (wood) *tchu* (who weeps)— that is, "weeping tree" in Peruvian.

The history of rubber might have ended there had it not been for the intervention of a small additive that changed everything and the persistence of a penniless ironmonger from Philadelphia.

CHARLES GOODYEAR: A HUMAN ADVENTURE

This is the West, sir. When the legend becomes fact, print the legend.

—John Ford, *The Man Who Shot Liberty Valance*, 1962

Goodyear knew little about latex. He attempted to make innovative valves for life preservers. However, in the early 1830s the demand for rubber diminished as suddenly as it had begun; customers abandoned the material, which in winter became as hard as wood and even shattered and in summer turned sticky and malodorous. Over the years Goodyear tried all manner of experiments to find a solution. None worked, and his financial situation suffered. His scientific luck, however, took a turn for the better. One day in 1839 Goodyear was boiling latex with sulfur between 140° and 200°C. Sulfur has chemical properties similar to those of oxygen. It, too, is capable of forming strong links between chains, but it is less reactive than oxygen. That is why it never sabotages its own work by cutting molecules, the reward for which is the durability of natural rubber. It took only a few links—a mild chemical reaction—to radically alter the state of the latex and transform it into a completely new, long-lasting material. One out of every two hundred latex carbon atoms reacts with sulfur. Two atoms of sulfur are arranged in such a way that they form a *disulfide bridge* between two chains. This represents a concentration of between 0.5 and 3 percent sulfur. If the concentration is too weak (say, just a trace), the link between the chains is insufficient, the rubber is not rubber, and the life of the resulting material is very nearly that of the latex used by the Indians. If the concentration is too strong, for example, 30 percent, the distance between the links is too short, and the network gets impossibly tangled. Too much sulfur kills the elasticity, and the solid product will not bend . . . but it is ebonite! One simply has to know what one wants.

In fact, all solids are elastic. Pulled, they will stretch in proportion to the externally applied force and then recover, but only if the amount of stretching is very small, on the order of 1 percent. Now, a rubber band can stretch five to

Top: molecular-scale image of a section of rubber at rest—a ball of long, disordered, tangled molecules connected by a few bridges. Bottom: the same section when it is stretched in the direction of the arrows and when the stretching is sustained. The molecules are always arranged without any particular positional order. Yet they present a preferential orientation: that of the direction of stretch.

ten times its length, and it is precisely this quality that distinguishes the material from other solids. The atoms are all separated by the same distance, d, in a crystal. Stretched *in the elastic domain*—within which there is no permanent damage to the material after withdrawal of the stress—the distances d of the crystal all undergo the same elongation. The atoms are no longer in equilibrium, and their interactions force them to draw closer together again. The mechanical action applied to deform the crystal tends to oppose these forces of attraction. When it is suppressed, the atoms find themselves once again separated by a distance d. Thus, separating the atoms, no matter how little, requires modifying the atomic structure.

This has something to do with the reason the elastic deformation of a crystalline solid is so restricted: The distance between the atoms will never reach ten times distance d. The elasticity of rubber is strictly tied to the existence of bridged molecular chains, and only a few bridges at that. Polymer scientists refer to cross-linking between molecular chains. A network of molecules is created, and communication between them is ensured by the transverse links between entangled chains. A chain folds over itself between two links that depend on two other chains. When it unfolds, the distance between these links is substantially increased without affecting the links between nearby atoms at all. The elastic force of rubber is fundamentally *kinetic*. It depends on the movements of a molecular chain. That is why an elastic strap heats on stretching and cools when it snaps back. Pulling on it causes the network of chains to orient themselves in the direction of elongation. When the pulling stops, the chains spontaneously return to their initial configuration. In the stretched state, their position is disordered (already present before stretching), and their orientation ordered. Simply put, the rubber is in a nematic liquid crystal state. This state is forced; it owes its existence purely to a mechanical constraint. Take away the mechanical constraint, and the nematic order disappears.

Described as an irascible inventor, Goodyear waited too long to apply for international patents for his invention. Hoping to expand his business in Europe, he sent samples of his sulfur-treated latex to England. The British rubber pioneer Thomas Hancock noticed that one of the samples had traces of yellow on the surface that left no doubt about the origin of the magic additive. In 1843 he reinvented the process of vulcanization and beat out Goodyear, who was just on the point of applying for an English patent. One of Hancock's friends is believed to have given the product a name related to Vulcan, the god of fire in Greco-Roman mythology, whose forge was situated in the shadow of Etna, in Sicily. Goodyear was plagued by money problems for the rest of his days and died deeply in debt in 1860. Although, in time, his family managed to live comfortably, Goodyear had no connection to the illustrious company that bears his name.

In Goodyear's day, scientists had no interest in polymers. In fact, they were not even aware that they consisted of long molecules. When chemists made them by mistake, they threw them down the sink because they equated them with badly behaved mixtures. A polymer has no well-defined melting point. In those days, that was the principal criterion for distinguishing a mixture, each component making a contribution to the melting point, which ranged widely. Scientists had no idea that, with polymers, melting did not happen all at once; the solid material softened over a fairly large range of temperatures before liquefying. Conversely, these materials refused to crystallize at low temperatures. Several decades passed before the special behavior of polymers began attracting attention rather than being considered a by-product of accidental chemistry.

The idea of long-chain molecules gradually took hold in the 1920s, thanks to the determination of a German chemist who went on to receive a Nobel prize in 1953 for his contribution to macromolecular chemistry. Hermann Staudinger successfully synthesized molecules with ever-longer chains by refining the control of the way in which they assemble. Consequently, he also came to understand the cause of the broadened melting point, which constituted a genuine paradigm shift in organic chemistry. Staudinger's career was beset with battles against received wisdom, such as that molecules were limited in size. Many chemists followed in Staudinger's footsteps, solving the structure of natural macromolecules and creating new ones, frequently inspired by natural materials. Examples include nylon, the synthetic analog of silk, which became wildly successful, and Kevlar. Plastic has been the most widely used synthetic material since the 1970s. Goodyear had already seen in rubber the versatility of modern plastics: He referred to *vegetable leather* and *elastic metal.* More than 70 percent of natural rubber comes from Southeast Asia and represents 42 percent of the total product (both natural and synthetic rubber). Tire production consumes 70 percent of the supply of natural rubber, to which are added between ten and twenty ingredients, each having its own function: antioxidants, softeners, colorants, and carbon black, which increases wear resistance.

Rubber is a fine example of practical invention preceding scientific understanding. It is also an illustration of a happy experimental accident ending in discovery.[1] Nonetheless, Goodyear always insisted that adding sulfur to latex had been a deliberate decision, without ever making his motive clear. He eluded the question by pointing out that an accident—hypothetically speaking—in his experimental approach would have ended in success one way or another because he would never have given up. In this, Goodyear presaged Louis Pasteur, who liked to say that, in science, chance favors the prepared mind.

BEYOND RUBBER

The Renaissance painter, hands on hips, stands before his canvas. Asked what he is doing, he answers, "I am waiting for it to dry," clearly annoyed at being interrupted in his artistic activity by such a banal question. We, mischievous travelers from the future, waggle an index finger: "You mean you are waiting for it to polymerize!" For we know that ambient oxygen is busy binding to the long molecules of linseed oil and, unlike with latex, at many more points. Before leaving, we tell the grumbling painter that he could cut his waiting time if he were to add a metallic oxide–base drying agent to his painting—cobalt, manganese, lead—which would accelerate polymerization by adding more oxygen. Iron oxide would be a poor choice because it does not give up its oxygen easily. According to manufacturers, drying agents should be used in only the smallest amounts. Painters already knew long ago not to exceed a tiny drop for each squeeze of paint to avoid premature cracking. Here, too: not too much, not too little. In the eighteenth century, one particular drying agent—also used as an eyewash and a treatment for gonorrhea in the king's hospitals—was white vitriol and contained . . . sulfur. Yet another (unexpected) link to rubber.

A curling iron makes curls. Keratin, the most abundant molecule in hair, contains sulfur atoms that create bridges within a single strand of hair and between adjacent hairs (through cross-linking). The outer layer of a strand of hair, the cuticle, contains the most keratin. Accordingly, a permanent is the result of modifying the chemical structure of hair: Hair is deformed, or many strands of hair (a curl) are stiffened by polymerization and cross-linking. Disulfide bridges form between the sulfur amino acids of the keratin. Hair straightening consists of reducing the disulfide bridges and then recombining them through oxidation.

Broadly speaking, the materials known as elastomers—from "elastic" and "polymer"—possess elastic qualities that are related to those of rubber; they are also cross-linked polymers.

Sulfur thus plays a role as a bridging agent between latex molecules. Imagine that you have just discovered the bridging agent between molecules of . . . water. I say "imagine" because, although there may still be much to discover about water, you will have to rethink a great many scientific ideas and theories that have long proved their worth in research laboratories and everyday life—no easy task. What might one expect from tossing this magical powder into all of the world's oceans, seas, waterways, and wastewater? The consequences of the cascading polymerization of these masses of water send shivers down one's spine, especially at a time when technology-related terrorist threats of all kinds are increasingly conceivable. Is it really absurd to think that researchers and engineers might entertain such speculations?

Not really, as the half-science, half–science fiction story of *polymerized water* will show.

POLYWATER: A CASE OF PATHOLOGICAL SCIENCE

Nikolai Fedyakin is a lone Russian chemist who studied the properties of water in the 1960s. Within very fine glass tubes, water showed a higher boiling point, 150°C, than the normal 100°C; a lower freezing point, −40°C instead of 0°C; and a syrupy viscosity fifteen times greater than normal. In Moscow, Boris Derjaguin, a major authority on colloids (solid particles that are finely dispersed in a liquid, to cite just one example), directed a laboratory on the physics of surfaces. At odds with the centralized nature of Soviet science, Derjaguin took up Fedyakin's experiments and set about reproducing the phenomenon. The results, published in Russian scientific journals, at first attracted little international attention. Then, in 1966, however, Derjaguin presented his findings on *abnormal water* at a conference in England, and the Brits and Americans pricked up their ears. Some were able to reproduce the experiments, but others could not.

A debate ensued, and hypotheses multiplied over this strange state of water, *polywater,* or *polymerized water,* whose behavior suggested that of water whose molecules were bridging irreversibly by forming long chains that then became attached to each other, all in complete violation of the laws of chemistry. Some scientists were disdainful of so much fuss in the face of a few hastily conducted experiments, but others trumpeted a new, fourth state of water (after ice, liquid, and vapor). The scientific community dithered for several years, and colleagues on the physics of liquids conference circuit at the time say that the community often split down the middle. Much time and money were spent trying to duplicate the experiments. For, if they were true, the stakes were considerable. With a pinch of polywater, a Dr. Strangelove could cause an ocean to slowly and inexorably be transformed into a medium totally unsuited to life. An American physicist warned that Earth would turn into Venus and that polywater was clearly the most dangerous material ever made by humankind. Polywater turned up in Kurt Vonnegut's iconoclastic novel *Cat's Cradle*,[2] in which the author subtly lampoons human stupidity in a mix of futurism and antiwar sentiment. Vonnegut describes the activities of a Caribbean sect that possesses a weapon of mass destruction called Ice-9, an alternative form of water that solidifies at the temperatures required for life. The novel was published in 1963; science fiction had anticipated science.

In the end it was obvious that the water used in the initial experiment was contaminated; the Soviets had been working with poor-quality tubes, and glass had seeped into the liquid in the form of small particles of silica, which had aggregated into a network. The phenomenon could be reproduced by introducing impurities. American biologist Denis Rousseau reported analogous results based on sweat . . . that he extracted from his T-shirt after a hotly contested baseball game on the campus of his laboratory! Whence the unusual viscosity and the altered freezing and evaporation temperatures. Albeit in good faith, the Soviets had made errors in the experimental protocol and their interpretation, and the defenders of polywater were forced to acknowledge it. For the majority of researchers, the matter ended in 1972. This episode is frequently evoked as an example of pathological science. Distinct from the practice of scientific fraud, pathological science is born of a psychological process in the course of which researchers get carried away by their beliefs and hopes and ignore rigorous, objective scientific methods. The telltale characteristic of a case of pathological science is a description of a phenomenon that occurs at the limit of detectability or that is only weakly statistically significant. Pathological science also figures largely in the thick files surrounding "cold fusion" and the "memory of water."

One of the functions of science is to correct itself. That takes time when its pathological version has gone beyond one or two researchers and spread to several laboratories—which was the case here. In the meantime, people were angry at the actions of members of a narrow-minded establishment, *official science,* working for special interests and little inclined to acknowledge that a discovery might be upending their certainty.

In 1983 Derjaguin wrote to the journal *Nature* to protest the handling of the affair and his role in a book about to be published.[3] Aside from this one-time polemic, it is interesting to read his ideas, especially coming as they did long after the furor. Derjaguin minimized the pathological character of the science in this affair and emphasized instead the positive consequences of the research on polywater, which, incidentally, inspired theoretical studies on the structure of water. He refuted the thesis of naïveté in explaining why scientists were willing to credit such far-fetched hypotheses: "It would be more correct to say that nature prepared a malicious surprise." Derjaguin noted ironically that, in any event, he had met with skepticism throughout his long career, including vis-à-vis work that was approved and built on by the entire community, though "only after more than 35 years." Duly noted. Still, one might have hoped for a less soft response from this noted scholar.

Leaving behind speculations about water that behaves like a polymer, let us turn our attention to the behavior of a polymer in water. A smidgen, of course: It is the little additive that changes everything that interests us.

CHAPTER 5

THE FIREFIGHTER'S JET STREAM:

REACH FOR THE SKY

The height of the stream of water from a firefighter's hose is limited by pressure and by water loss. It would be annoying were it to reach only to the seventh floor when the fire is raging above. However, if the firefighter adds a tenth of a gram of polymer per liter of water, the height will be increased by 50 percent and the risk of the stream breaking up will be minimized.

Polyoxyethylene, or polyox, is a polymer that has very long and flexible molecules and is soluble in water. The increased height of the jet results from the additive's remarkable ability to reduce the friction of the water against the inside wall of the hose. This phenomenon, discovered in 1949, is called *drag reduction*. Intensive research efforts have been under way ever since, but there is still no explanation acceptable to all because much of the behavior of polyox in water is described qualitatively and remains controversial. The extremely dilute character of the solutions (only a trace of the polymer is necessary) is striking and counterintuitive; it does not help that the nature of turbulent flow also remains elusive. However, many hypotheses refer to a change in the nature of turbulence to explain drag reduction.

Polyox molecules, which form a randomly tangled ball in water at rest, relax once the fluid begins flowing. The unfolding of the chains mitigates the effect of the turbulence in the water. Turbulence occurs in particular near a pipe wall, and the associated friction increases with the velocity of the fluid. But this turbulence has no effect on flow; on the contrary, it is loss of energy that limits the height the water will reach. A polyox solution flows farther and more easily because the loss of energy is reduced.

Aside from the quantity of the additive, the effectiveness of the experiment depends very much on the polymer's molecular mass. The greater it is, the greater the drag reduction. The property of polyox is thus that of long and flexible polymers. However, the polymers must also be able to stretch; those with high molecular mass but no linear structure are ineffective. Such is the case with gum arabic, whose molecules are highly branched.

The energy necessary to pump drilling fluid is considerably reduced by the addition of polyox-type polymers. This effect was widely used even before explanations and theories began to disseminate. The addition to water of just a few parts of polyox per million[1] gives the same flow as with water alone but reduces the needed pumping pressure by 80 percent. Conversely, at the same pumping pressure, flow

Newtonian or Not?

The behavior of water and of alcohol is well known because these fluids are made of small molecules with simple internal structures. Their viscosity (a measure of the resistance of a material to deformation) depends on temperature and pressure but not on the way in which the liquids are made to flow (hard or soft, at length or briefly). They are called Newtonian fluids because their viscosity is related to the force applied according to a simple law established by Newton. Other fluids have non-Newtonian behavior—egg white, ketchup, yogurt, paint, and long, flexible polymers such as polyox in solution—because they are composed entirely or in part of large molecules that tend to adopt complex configurations. Their viscosity thus depends very strongly on the way in which the liquid is deformed, which obviously complicates the description of its behavior.

increases by 40 percent. Polyox is cheap—around fifteen dollars per kilogram—which is important in improving old sewage systems and in eliminating the need for reconstruction. In England, the city of Bristol is a case in point.

Sulfur and polyox were components added to rubber and the firefighter's jet stream. However, matter may also naturally contain a little additive whose mission is to endow it with a remarkable property . . .

CHAPTER 6

THE GLAMOROUS AFFAIR
OF GAS AND LIQUID

The stability of its foam is the trademark of a good champagne. It is due to the natural presence of amphiphilic molecules in the grape, amounting to just a few tens of milligrams per liter, positioned at the interface between liquid and gas. Most are proteins and polysaccharides. For *crème chantilly* and *mousse au chocolat,* the molecules involved are milk and egg white proteins, respectively. How can such a little quantity have such a big effect?

FROM GRAPE JUICE TO A CHAMPAGNE GLASS:
THIS WAY, PLEASE!

Freshly pressed grape juice—"must" in the parlance of wine making—goes through an initial fermentation in a ventilated vat. The result is a noneffervescent wine—the base wine—that is very acidic and tastes raw and unfinished. Because its foaming properties are a fair prediction of those of the ultimate champagne, laboratory experiments are generally done on the base wine, which is easier to manipulate. Champagne contains carbon dioxide (CO_2), which is present after a second alcoholic fermentation, in a sealed bottle that is positioned upside down and turned regularly to detach the fermentation sediment from the sides of the bottle. In the final step the neck of the bottle is frozen and the sediment removed by disgorgement. Although the "invention" of champagne may have been a collective enterprise, this process of double fermentation—the "champagne method"—is credited to Dom Pérignon (1638–1715), a Benedictine monk at the monastery of Hautvillers, near Épernay. A drawing liquor, which consists of yeast and sugar, is added beforehand to promote the second fermentation, called *prise de mousse* (formation of bubbles), for five to seven weeks. That's it. Next the wine is aged for a minimum of fifteen months for ordinary champagne and up to seven years or more for the finest selections, known as *cuvées millésimées*.

However, these processes have nothing to do with the spontaneous formation of bubbles in the glass. What causes those are the cellulose fibers from the dish towel used to dry the glass or microcrystals in the wine. The CO_2 that is present in the champagne escapes from the liquid and aggregates at these bubble-forming sites. Some glassmakers also make champagne flutes with an etched inner surface. A single fiber can be the source of thirty bubbles per second, which drift up with the regularity of a metronome. The result is the constant mixing of the champagne, which in turn drives volatile odor components to the surface of the beverage, liberating the flavor.

What is the secret of a beautiful, stable foam? How does it behave? These are fundamental questions that continue to fuel research in the field. Formulating a clear and precise answer is a difficult proposition because wines are natural products whose qualities depend on many uncontrollable factors, such as sunshine and rainfall. All base wines contain essentially the same concentration of alcohol and degree of acidity. It is thus other constituents, which are present in very small amounts, that actually explain the marked difference in the flavor and character of the foam. Consequently, it can vary significantly from one vineyard to another, one harvest to another, and even from one bottle to another.

THE ROLE OF INTERFACES

In pure liquid, a foam subsides immediately. Think of the foam caused by ocean waves. Bubbles cannot spontaneously arise in a liquid whose molecules are strongly linked one to each other. Making bubbles requires a liquid to share a large surface with air, which the liquid hates because its molecules prefer to be surrounded by others of their own kind. Accordingly, bubbling implies overcoming the forces of cohesion within the liquid, which takes energy. Foam results when the amphiphilic molecules stabilize the interfaces between liquid and air. These amphiphiles belong to a large family of *tensioactive* molecules, so called because they have the power to change the surface tension of a liquid. A commonly used synonym for tensioactive is *surfactant,* for *surface active agent.*

Surface tension arises at the interface of two immiscible fluids, such as air and liquid. To understand its origin, let us question a molecule inside the liquid, then one of its colleagues on the surface. The first molecule tells us that it is surrounded by neighbors for which it feels exactly the same pull: The total electrical attraction exerted on this molecule is thus zero. The cohesive energy in the bulk of the water corresponds to what it would take to separate the molecules. The second molecule, on the other hand, tells us something completely different: "Put yourself in my place: I have nearly half the number of friends as my colleague, which makes me very tense! In fact, I feel the force of attraction of my neighbors below, which, for me, is not zero and is directed toward the interior of the liquid, following a direction perpendicular to the surface." This force is the cause of the surface tension of the liquid, which forms a very thin skin, elastic and stretched, between the air and the liquid.

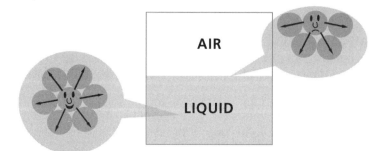

Left: a happy molecule. Inside the fluid, it is completely surrounded by its colleagues, with which it interacts: the forces of attraction cancel each other out in pairs (each arrow corresponds to another in the opposite direction). Right: an unhappy molecule at the liquid-air interface. Because the molecule is deprived of some of its neighbors, the sum of the forces of attraction between them is not zero.

At the interface with air the cohesive energy is less because a molecule there has fewer neighbors than a molecule in bulk water. The smaller the surface, the more favorable the cohesive energy—and each system wants to lessen its energy. Accordingly, the liquid molecules seek to diminish their surface, which creates tension. Surface tension is what causes water to bulge from the sides of a glass filled to the brim. It is also what enables featherweight insects like water striders to walk on ponds even though they are denser than the water. Their long feet are equipped with hydrophobic hairs that aid their flotation. Lay a piece of absorbent paper on the surface of a glass of water and quickly put a sewing needle on top of it. When the paper is sufficiently wet, it will sink to the bottom of the glass, whereas the needle—thanks to surface tension—remains floating on the water. Look closer: You will see that the "water skin" sags under the needle as if it were lying on a stretched piece of plastic film. The needle causes the skin to curve into the interior, increasing its surface, which induces a force directed upward that compensates for the weight of the needle—akin to the situation of a gymnast on a trampoline. This force is attempting to restore the planarity of the liquid skin.

CERTIFIED FOAMABILITY

Surface tension has a decided influence on a liquid's foamability. It is extremely sensitive to the presence of surfactants at a liquid-gas interface, which they stiffen, enabling temporary encapsulation of the gas. Amphiphilic molecules spread out to the interfaces: Their hydrophobic tails are exposed to air, and their hydrophilic heads are in contact with the aqueous medium. The role of these molecules is not to get rid of surface tension but to reduce it. Why? The hydrophobic tails at the surface do not like being piled up. They repel each other, which creates a force that opposes the surface tension of the water.

The concentration of surfactant in champagne is necessarily less than that at which the additive cannot dissolve (known as the *limit of solubility*). Otherwise, the amphiphilic molecules would precipitate in the liquid before having accomplished their creative activity. If the quantity of surfactant is too low, the foam is mediocre. For aside from the molecules needing to have the time to propagate through the mix to the interfaces, there also must be enough of them to form a dense, resilient layer. A small change in the quan-

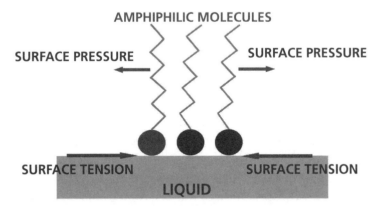

Several amphiphilic molecules on the surface of a liquid. The molecules of the liquid try to minimize its surface, which gives rise to surface tension. For their part, the amphiphilic molecules do not like being crowded, and they repel each other. The result is tension forces in the opposite direction due to *surface pressure*. The upshot of the contest is the reduction of the liquid's surface tension.

tity of surfactant or the presence of a few impurities is enough to alter the formation and stability of the foam. This sensitivity enormously complicates the study of foams. A foam represents a large volume produced from a small amount of matter. This fact, together with the undeniable observation that the stabilization of interfaces takes place at the molecular level, requires us to acknowledge that it takes a ridiculously small amount of surfactant to alter the behavior of a gas-liquid mix—a soap bubble, for example, typically contains one amphiphilic molecule per one thousand molecules of water—and that a flat wine can give rise to a sparkling drink.

LIVING ON THE EDGE

What causes a stable foam to collapse? Foamability—a measure of the volume of foam formed from a volume of liquid—is distinct from the stability of the foam, which is characterized by its lifetime. Foam may collapse for three different reasons. First, it immediately breaks down in the absence of a surfactant (or when there is too little surfactant). Second, drainage of the liquid component of the foam toward the bottom, related to weight, dries out a foam and can happen very quickly; it is faster when the bubbles are large because they deform easily. Finally, aging due to differences of pressure in the bubbles can occur on a timescale of minutes to hours.

Gas pressure varies among bubbles; the smaller the bubble, the greater the gas pressure. When gas diffuses from one cavity to another across the network of alveolated film, the smaller bubbles lose their gas to the profit of the larger ones. They grow smaller and smaller and finally disappear. The effect of drainage is accentuated when the film becomes increasingly thinner, and the bubbles ultimately pop. But foams resist aging. When interfaces poor in surfactant form (such is the case when an alveolated film is stretched), the surfactant molecules move to the sparsely populated site in a beneficial process of self-healing. In an alcoholic beverage in which the dissolved gas resulting from fermentation is CO_2, which is soluble in water, an aging foam collapses in a few minutes. Some producers prolong the life of their beer—like Irish brown stout—by introducing nitrogen, which is less soluble in liquid than CO_2 and promotes separation of the gas and liquid phases. The bubbles of this foam—called wet—are smaller and keep their spherical form longer: Gravity makes the liquid harder to extract, and the foam is less susceptible to deformation.

BEYOND CHAMPAGNE

The principle underpinning all foams is the same: Gas is dispersed in a liquid or paste, and the interfaces created by this phase separation are stabilized by a natural or an artificial surfactant additive.

Detergent surfactants help to get rid of dirt, but there is no direct connection between the effectiveness of the washing and the foaming power. On the contrary, too much foam is counterproductive: It reduces the force of the water on the plates in a dishwasher and results in less effective cleaning. *Ghassoul* (an Arabic word) is a material that washes hair perfectly well without foam; added to water, this powdered clay produces a soft paste that, like blotting paper, absorbs impurities and oil.

The mining industry is the biggest consumer of foam. In flotation vats, bubbles attach to minerals, separating them from the gangue and allowing them to float to the surface.

In processing textile fibers (such as waterproofing or carpet cleaning), the covering power of a foam helps to spread the product, dramatically reducing the amount needed.

Foam spread on a pollutant (a radioactive substance, say, or an oil spill) first neutralizes it, then allows it to move; here the trick is to ensure that the attraction between the contaminant and the foam is greater than that between the contaminant and the substrate to be cleaned. Decontaminating a dirty foam involves treating a smaller amount than would be needed if flushing with water, which is all for the better. By chemically reacting with the product to be removed, the surfactants used in cleaning a surface work more effectively when they are dispersed in a foam than in a liquid.

To combat fires, a foam that acts locally, thereby limiting water damage, is very desirable. It blocks the oxygen feeding the flames. In a confined environment, it can be lifesaving for people trapped in a conflagration.

Some foams can be solidified by heating them. In a metallic foam, once the metal powder has melted, the foaming agent introduced to form the alveoli disappears as a gas, and what is left after cooling is a solid, light material. In an automobile chassis, metallic foam absorbs shocks and acoustic waves (cracks and sound propagate poorly in a cellular structure where air and dense material alternate) and thermally isolate and lower the mass of the engine, which reduces gas consumption. The high cost of producing metallic foams still limits their use. For bread or cake, heat helps the dough both to rise (it is nothing more than . . . foam) and to become firm. Egg whites beaten into snowy peaks are essentially air, proteins, and water. To obtain more foam with a single egg white, since air is plentiful, add more protein or water or both. It is easier to add water in a steady stream while beating well. In this way one can produce up to a cubic meter of foam with a single egg white and, through heating—after adding a little flavoring (for taste!)—liters of very light meringues that Hervé This and chef Pierre Gagnaire very prettily call *wind crystals.*

The formulation of foam continues to be a matter more of savoir faire than of the application of a theory that still resists taking into account the molecular role of surfactants.

However, even if foams are ubiquitous in everyday life and widely used in industry, one also needs to know how to destroy them—and rapidly . . .

CHAPTER 7

DOWN WITH FOAM!

The stomach secretes foam as a result of excess acid; for a drug to get through, it must therefore be taken with an antifoam. Processing wastewater requires breaking down mountains of honeycombed films created by tons of detergent. During fermentation, the formation of bubbles is driven by air and promoted by microorganisms that produce surface-active proteins; this secondary effect may not be desirable. Antifoam industries also include oil, wood pulp for paper, sugar, water-based paints, and plant processing.

Both antifoamers and defoamers are made to be spread over a foam. It takes only the smallest amount of either. How do they work? We do not fully understand the process, but some things are known. A defoamer consists of less than 5 percent hydrophobic particles, smaller than a micrometer, dispersed in oil or a mixture of oils of different viscosities to optimize performance. A wide variety of chemistries is possible depending on the nature of the foam to be eliminated and the environment that gave rise to it—a natural habitat to be protected from additional pollution by defoaming or a closed tank in a factory. Oil alone is capable of destroying foam, but its effectiveness is increased with the addition of hydrophobic molecules: Their role is similar to that

PARTICLES SUSPENDED IN OIL

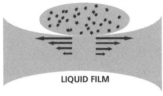

LIQUID FILM

a

HYDROPHOBIC BRIDGE

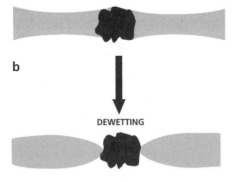

b

DEWETTING

c

Defoaming in three steps. (a) Cross-sectional view: a drop of oil loaded with hydrophobic particles reaches the liquid-air interface. The liquid film below drains—escapes!—depicted by the arrows. The film becomes thinner. (b) Zoom view: the thickness of the film is now roughly that of the particle. A hydrophobic bridge is created between the two edges of the film. (c) Liquid does not like hydrophobic substances: it seeks to escape the surface of the particle through a phenomenon known as *dewetting,* and the wall of the foam collapses.

of a needle piercing a membrane. Vaporized into droplets measuring around ten micrometers in size, the particle-charged oil (which is insoluble in liquid) spreads to the interfaces and thins the honeycomb film by draining the underlying fluid. Chemists take into account the wettability of the drop of oil at the interface according to the desired mode of action: Will it spread or bead up? Defoamers are hydrophobic; otherwise, the droplets would disperse in the aqueous medium and would not contribute to the dewetting action of the defoamer. Once drainage has reduced the thickness of the liquid films roughly to the size of the particles, the particles form hydrophobic bridges between pairs of bubbles. The bridges trigger dewetting at the hydrophobic surface of the particles, ultimately resulting in a hole and the catastrophic breakdown of the film. Pigments in lipstick destroy champagne froth, and a second shampoo foams more than the first (the dirt particles have been removed).

The weak foaming behavior of whole milk, compared with that of skim milk, and the destructive effect of a small quantity of egg yolk in beating egg whites are examples of the destabilizing effect of fats. Beware of plastic containers, for fat attaches to them. Even a few traces of egg yolk in egg white, say, 0.03 percent, will prevent foaming. Lecithin molecules bind to the long proteins of the egg whites and impede their assembly into a network. Moreover, the hydrophobic parts of the fats of yolks bind to the proteins of the whites, which reduces their availability for coating air bubbles.

We can now leave the world of additives that harmoniously link gas with liquid, satisfied with having described this goal through the decrease of surface tension and the stabilization of liquid-gas interfaces. That is already an achievement. However, it would be interesting to apply the same ideas in a radically different context, one related to the fundamental processes of life: allowing oxygen to pass through blood. No more, no less.

CHAPTER 8

BREATHING:

AN UNSEEN TRIUMPH

WHAT IS THE PROBLEM?

The lung is alveolar tissue covered with a film of liquid composed of 90 percent water and 10 percent mineral salts, phospholipids, and proteins. The alveolus, the functional unit of the lung, is a tiny elastic sac that permits oxygen to pass into the blood. Lungs are full of alveoli: three hundred million for an adult, or a surface that is between seventy and two hundred square meters for a tissue thickness of only a few micrometers. Breathing increases pulmonary volume by opening the thoracic cage and contracting the diaphragm, which increases the air-exchange surface area of the alveolar film by ten square meters. For the transfer to be efficient, the surface must be as large as possible. However, it so happens that the natural tendency of surface forces is to cause the alveoli to close or contract, at the risk of collapse, because surface tension makes it hard for them to open again.[1] How is it possible for every human to engage in this battle twenty thousand times a day, unawares, and emerge victorious?

THE HISTORY OF ALVEOLAR PRESSURE

A cluster of alveoli looks like a bunch of polyhedra, or rows of honeycomb, as a result of which two alveoli share a single membrane. To estimate the pressure[2] on both sides of an alveolus, we can compare an air sac surrounded by physiological liquid to a bubble of air in a glass of water. The surface tension of the air-water interface compresses the bubble to reduce its surface, which increases the pressure inside the bubble. When the air pressure on the inside is equal to the sum of the outside pressure, exerted by the water, and that exerted by the interface, the bubble is said to be in equilibrium. *Interfacial pressure* is governed by a law formulated by physicist, mathematician, and astronomer Pierre-Simon Laplace (1749–1827). It is so simple to express as an equation that it would be a sin not to supply it here. The law states that the difference in pressure of any part of the membrane of a spherical bubble in water is equal to twice the surface tension divided by the radius of the bubble:

$$
\begin{aligned}
\text{Interfacial pressure} &= \text{internal pressure} \\
&\quad - \text{external pressure} \\
&= 2 \times \text{surface tension/radius}
\end{aligned}
$$

The smaller the radius, the greater the difference between the internal and external pressure. For a bubble of one millimeter diameter, the difference is on the order of

0.3 percent of atmospheric pressure. If the diameter is a thousand times smaller, the difference is a thousand times greater, or three times atmospheric pressure (0.3 percent × 1,000 = 3). Our alveoli are in direct contact with air: The air pressure flowing into the alveoli is equal to the atmospheric pressure. Laplace's law becomes:

$$\text{External pressure} = \text{atmospheric pressure} - 2 \times \text{surface tension/radius}$$

The external pressure on an alveolus is the pressure of the surrounding fluid. According to this equation, external pressure is less than atmospheric pressure. If the liquid is pure water, solving the equation will show that the pressure on any part of the alveolar membrane is equal to 1 percent of the atmospheric pressure. This does not sound like much—a tiny percentage—but in the context of breathing it amounts to having several kilograms weighing on the chest. Moreover, the fact that the alveoli vary greatly in size complicates the situation, for Laplace makes it clear that pressure is higher in little cavities than in large ones. If two alveoli of different sizes are connected, the air from the smaller one will in fact migrate to the larger one. Breathing thus becomes a heroic act! If this activity is ultimately mechanical, it is because some of the cells in the pulmonary membrane synthesize a surfactant in a carefully regulated quantity. These cells, called *type II pneumocytes,* take up only a tiny percentage of the alveolar surface. The surfactant in question, a mix of phospholipids and lipoproteins, lowers the surface tension at the air-liquid interface . . . just as the surfactant in champagne does. It divides it by three, which similarly reduces the amount of work. Now, the pressure among the different-sized alveoli is balanced. During the course of the respiratory cycle, new surfactant molecules are produced so that the quantity inside each alveolus remains the same, for secretion is invariably lost during compression and expansion. The size of an alveolus enlarges during inhalation and diminishes during exhalation. "Breathe in! Breathe out!" exhorts the physician. "I'm increasing surface tension! I'm decreasing it!" the surfactant responds. On inhalation, the surfactant molecules spread over the surface. On exhalation, clustered together and compressed, they form a multilayered structure, a veritable natural liquid crystal. This is a two-step mechanism, a terrific self-regulator that enables equilibration of pressure throughout the respiratory cycle.

NOT TOO MUCH, NOT TOO LITTLE

Respiration itself favors the production of surfactant. The deep breaths that we sometimes make without thinking may serve to regulate that production. A shot of adrenaline, which accelerates respiratory rhythm, similarly stimulates surfactant release. If there is too much surfactant, the alveoli swell and may even collapse: The gaseous exchange with blood is disturbed, and the respiratory muscles exhaust themselves trying to reopen the air cells at each inhalation. Respiratory distress can also be caused by a deficit of surfactant because the alveoli gradually shut down. This disorder typically strikes premature infants younger than thirty-five weeks of age because surfactant is slowly synthesized only during the second half of pregnancy. When distress occurs, the pediatrician will have the baby inhale an aerosol containing the precious surfactant.

PULMONARY SURFACTANT:
A MULTIFUNCTIONAL MATERIAL

Pulmonary surfactant does not just promote the opening of alveoli during inhalation and prevent their complete closing after exhalation. It also lubricates and serves as an antiadhesive. There is nothing surprising about finding a natural surfactant in tissues where surfaces must slide over one another without sticking, as in the pleura (a membrane protecting each lung), the pericardium (a layered saclike tissue enclosing the heart), and the membranes of the joints.

Pulmonary surfactant prevents blood salts and proteins from escaping toward the alveoli. If the quantity of surfactant is insufficient, pulmonary edema may result. Too much water and mineral salts will cause the alveoli to swell. The patient coughs, spits, and may ultimately suffocate.

Pulmonary surfactant facilitates the flow of mucus, the viscous liquid that protects the respiratory system by trapping and exporting the dust contained in the air we breathe.

Finally, pulmonary surfactant acts as a bactericide. Its proteins carry sugars that coat bacteria, dooming them to phagocytosis and degradation by macrophage cells.

Biologists continue to explore the mechanisms that control the release of pulmonary surfactant and to research alternative products for treating its deficiency.

CHAPTER 9

FAMILIARITY AND DISTANCE:

COLLOIDS

Liquid has found another love interest! This time it is granular matter.

BEING DIVIDED BRINGS THEM TOGETHER

Solid grains suspended in liquid make ink. If they are borne by gas, the result is smoke. Liquid dispersed in gas is aerosol. Adding air to a liquid creates foam. Butter is a dispersion of water droplets in fat. Blood is a liquid—plasma—swimming with red and white cells and platelets. An emulsion is one liquid dispersed in another.

What connects all of these materials is that they contain finely divided matter. Whence *colloid*,[1] a substance made of particles "of colloidal size" in the solid, liquid, or gaseous state, dispersed in a continuous medium whose phase or composition is different. As for the idea of colloidal size, in 1972 the International Union of Pure and Applied Chemistry saw fit to reserve the term "colloid" for objects with at least one dimension that measures between one and one thousand nanometers. In practice, however, the definition is far less restrictive. In fact, the colloidal domain is associated with a length scale somewhere between atoms (or "small" molecules) and objects observable with an optical microscope.

Transition from the bulk state to the colloidal state is accompanied by a large increase in the relationship between surface and volume. A single gram of particles ten or so nanometers in diameter can develop a surface—the sum of the surfaces of all of the particles—of several hundred square meters.[2] One senses the impact of this surface on the properties of colloidal material.

THE SCRIBE'S TASK

The scribe who set about making ink to write on papyrus had a sudden insight: Dispersing grains of charcoal in water and shaking up the mixture is easy, but it remains stable for only a few hours. After that the grains aggregate, and the rigid objects that result sink to the bottom of the vial under the effect of gravity. It is only when the particles are sufficiently small that thermal agitation can combat sedimentation, or the opposite: *creaming*—when the particles rise because their density is less than that of the fluid.[3] The scribe's desire was to stabilize the dispersion of granules in liquid. He finally achieved this objective, which is still relevant for many of our industries today, using sap from the acacia tree. Once again, invention preceded explanation.

MÉNAGE À TROIS

How do you promote the long-term cohabitation of a liquid-particle couple? By organizing a ménage à trois. The guest is called a polymer. The polymer in acacia sap is gum arabic.

Presented this way, the arrival of the polymer may seem an intrusion. However, we can already see that it is not a bothersome guest as it constitutes only a tiny quantity of

SEVERAL THOUSANDTHS
OF A MILLIMETER

SEVERAL MILLIONTHS
OF A MILLIMETER

The three members of the colloidal family: liquid, grain, and polymer.

Left: grain-liquid couples—divorce. Right: mix of grains, liquid, and 0.1 percent polymer grafted onto grains—a successful *entente à trois.*

the solute, less than half a percent. Although discreet, it dramatically alters the life of the ménage. Particles in suspension remain in proximity, all the while maintaining an individuality that prevents them from sticking together. To ensure this colloidal stabilization, the long molecules of the polymer must attach to the particles. The polymer-guest molecule consists of two parts: One has a strong affinity for the liquid, whereas the other cannot abide it and prefers to attach itself to the particle. When two attached particles meet, the interpenetration of the polymer chains each supports is limited; consequently, the particles, although still susceptible to the forces of attraction, keep their distance.

NEITHER TOO LITTLE NOR TOO MUCH

If there are too few polymer molecules, the surfaces of the grains are starved, reduced to sharing the rare, isolated molecules. A little like vacationers who arrive early at the beach, these do not interact and have plenty of space to lie down. The grains in turn preserve their mutual affinities, forming bridges: Sedimentation resumes.

When the surfaces are sated, the polymer molecules form brushes because they no longer have enough space to stretch out (the beach gets increasingly crowded as more vacationers arrive), and the brushes prevent the granules from getting too close; their dispersion is ensured.

Too much polymer is the enemy of the good: Introduced in excess and deprived of space on the surface, the free polymer molecules move toward the brushes already attached to the granules and get all tangled up. The colloid is transformed into a gelatinous substance with undesirable properties.

The stability of the colloid is thus the result of a fine dose of polymer in a solute. Colloidal stabilization by a polymer is called *steric stabilization* by loose analogy to what in chemistry is called *steric hindrance*—a state in which the volume occupied by a part of a molecule blocks the approach of a reactant or another part of the molecule. This analogy is deceiving: The real reason polymer molecules prevent particles from aggregating in a colloidal fluid is that they retain liquid.

ADDITIONAL INSTRUCTIONS

Indeed, the success of the ménage à trois is due to the liquid properties. In a "bad" liquid, the contacts between polymer chains on the one hand and molecules of liquid on the other are favored to the detriment of the molecule-chain contacts; whence a tendency of the chains to interact and to expel liquid: guaranteed incompatibility. In contrast, in a "good" liquid, mixed contacts are promoted, and the chains mutually repel each other: Particle dispersion prevails.

ANTITHESIS

In certain situations, provoking phase separation is desired in order to extract solid elements. Coagulants such as aluminum salt are used to treat wastewater. Seeing that the wine has become cloudy as a result of unintentional fermentation, a winegrower dilutes an egg white in water; this *wine pasting* encourages the aggregation of particles in suspension, which enables the "fining" and decanting of unclouded wine. Once upon a time there were bacteria in aquatic environments that needed to be able to see light in murky water. They began to secrete polysaccharides—those long, natural polymers we imbibed with our champagne—that caused the particles to coagulate and to drop to the bottom of the pond. And the bacteria saw light! Over the course of time, the bacteria contributed to the autopurification of water.

POSTSCRIPT

A clay such as kaolin is a colloidal suspension of mineral particles in the form of platelets several micrometers square and dispersed in water. However, no polymer is grafted onto these sheets: Their stability is ensured by the spontaneous presence of electrical charges all around, the natural by-product of the chemical structure of kaolinite. The stabilization is the product of a subtle equilibrium of the forces of repulsion between the layers of charges carried by the platelets and the forces of attraction between the platelets.

Milk is an emulsion of droplets of fat in an aqueous medium stabilized by proteins.[4] Around 1560, surgeon Ambroise Paré realized that liquids like milk and cream are composed of water and fat; he coined the term "emulsion" based on the Latin *emulgere*—to milk. Milk contains two principal proteins: casein and another that derives from lactoserum (whey), a yellowish liquid that results from coagulation. The stabilization of the emulsion is reinforced by the casein molecules, which, arranged on the surface of the fat droplets, guarantee repulsion between micelles because the casein molecules are negatively charged. However, this electrostatic repulsion is not enough to prevent the occasional coalescence of micelles; because they are larger, they rise to the surface (fat is less dense than water) and form cream. The effect is even faster when milk is heated because the micelles move more rapidly, and collisions, which promote fusion, are facilitated. Beyond 80°C, casein coagulates and forms a skin on the surface of the pan of milk. Coagulation also occurs when a little salt (or acid: vinegar or lemon) is added to the milk because the positive ions of the salt are attracted by the negative charge of the casein. They neutralize the forces of repulsion between the micelles, which merge.

Electrostatic stabilization is an alternative to that offered by polymer grafting, but the polymers have a clear advantage for industry: They do not need to be as finely regulated as a concentration of ions, they are suited to nonaqueous solutions, and they work for a large range of particle concentrations, whereas stabilization by charge is effective only for weak concentrations.

Coming closer a little farther away . . . that is one way to describe the success of the colloidal ménage. And may its peace and harmony spread to the kitchen! Off we go to take a look at edible colloids.

CHAPTER 10

SENSITIVE COOKING

> The daily special is a great idea so long as you know what
> day it was prepared.
>
> —Pierre Dac

BARCELONA, CARRER DEL CARME

"Don't be deceived: My job is 3 percent investigation and 97 percent creativity. And if I asked myself why a recipe works, I would go crazy!" says Ferran Adrià, ushering us into his lair in Barcelona.[1] In the quest for sensitive matter, nothing beats getting into the kitchen when it comes to applying the concept to cooking. So naturally we decided to go see the most radical chef of our time, head of El Bulli, on the Costa Brava. "From April to October, each evening, sixty-five 'actors' gather at the restaurant for a mere fifty 'spectators,'" he says wryly. The rest of the time, the high priest of molecular gastronomy and his team labor in his workshop in Barcelona to create the recipes for the coming season. El Bulli was voted the world's best restaurant for 2006, 2007,

2008, and 2009 by the British magazine *Restaurant*, the "go-to" publication on the subject, which each year asks a panel of five hundred critics, chefs, and gourmets to put together a list of the five hundred best eating establishments in the world. Adrià is a futurist cook. He experiments with new textures and flavors, fusion, dehydration, and cryogenics, mixing heat with cold, solid with gas, and crunch with sparkle. Like Mandarin air with bitter coconut, parmesan spaghetti, and truffle ravioli à la carbonara. To surprise and to confuse are his hallmarks.

"Sensitivity is everywhere in cooking!" he says by way of answering our eager request for examples. "A knife can totally change a material: An apple cut into fine slices is very different from an apple cut into thick ones." A disarmingly simple observation. What preparation might be irreversibly altered by a very small variation in temperature? "Which one isn't? For humans, isn't life or death a difference of just three degrees?" His practical wisdom is compelling:

> I would add all the same that you'll find the most science in pastry, where everything is very measured. Normal cooking is more empirical, more a matter of intuition. To put it more precisely, I would say that there is no practice more rigorous than that of the pastry chef.

His next cooking challenge? The question goes unanswered. Adrià tells us that what he really wants at present is to lessen the amount of time he spends on the business part of things to return to his real métier and explore new flavors and ways of combining ingredients from the four corners of the globe, the Amazon in particular. "When you think that nature produces two thousand five hundred sorts of citrus fruits!" Still, from the doorway as we leave, Adrià volunteers a little impromptu advice: "Don't forget to go check out agar-agar!" He is right: This extract of red algae, which goes under the name of E406 on package labels, is used in desserts, creams, sauces, and jellies for its amazing gelling ability. A few grams suffices to gel a liter of liquid after a minute of boiling. The effect is reversible: At high temperatures, the gel reverts to a liquid. Agar-agar has been an essential ingredient of Japanese cooking for a thousand years. Gladly, then, will we follow its lead to sensitive cooking through the first of a series of four doors, on which our objectives are clearly labeled: *contain, stabilize, coat, thicken.*

CONTAIN

To be lifted with a spoon or with the fingers, a gel must maintain its form, but afterward it must flow. We expect it to retain its liquid components under normal conditions of storage but to liberate them when a mechanical force is applied. Yogurt, gelatin, fruits and their derivatives (jellies), meats and their derivatives (cooled sauce from a roast), and cosmetic and pharmaceutical creams are liquid substances—water, fat, emulsion—whose flowing nature is restrained by the addition of a small amount of a gelling agent. Gelling agents are macromolecules whose origin may be biological (gelatin, polysaccharide, protein) or synthetic (polyacrylate, polyacrylamide). These macromolecules create a network of pathways that propagate through a liquid. The elasticity of the gel depends on the number of pathways. By increasing the concentration of the gelling agent, the gel's consistency can very easily go from weak, as in a cream, to very hard, as in a paste. A gel is Janus faced: It is "liquid" on the level of "small" molecules or ions diffusing freely in the medium and "solid" at the macroscopic level because the network formed by the gelling agent is capable of absorbing the mechanical energy applied during deformation.

In the case of agar-agar, the story comes down to an aqueous phase that is fluid at high temperatures but gels on cooling, around 40°C. The same behavior holds for other plant polysaccharides (alginate, carrageenan, agarose, amylase, pectin [for jelly]) or bacterial polysaccharides (xanthan, gellan) or even gelatin,[2] obtained from the collagen of animal tissues (skin, tendon, cartilage, bone), as well as other proteins related to gelatin whose native structure is helical. At high temperatures, the macromolecules of these gelling agents appear as disordered chains dispersed as coils in the liquid without associating. Lowering the temperature promotes new interactions (hydrogen bonds) between the chains, which adopt helical configurations that are more rigid than the coils: This is the *coil-to-helix transition.*

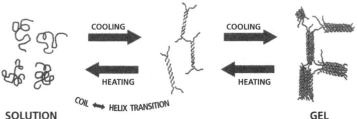

How do the macromolecules of a gelling agent organize themselves during the transition from a fluid solution to a gel state and vice versa? At high temperatures, the macromolecules are tangled up in coils in the liquid and ignore each other. When the mix cools, the interactions between the macromolecules are altered: they prefer to associate in helices at an initial temperature. If the concentration of macromolecules is sufficient, another transition takes place at a second temperature: a network of tangled helical strands propagates throughout the liquid. Gelling has occurred.

Subsequently, as the temperature continues to decrease, a second transition—depending on the concentration of macromolecules and their mass—occurs: The helices link up to form strands. Finally, the solution becomes a gel when the network of strands propagates throughout the liquid.

The opposite transition is produced by raising the temperature, but the characteristics are very different: The melting temperature of the strands is several tens of degrees higher because their helices are stabilized by the bonds between neighboring macromolecules,[3] and energy released as the bonds break—not present at the transformation of solution into gel (known as the *sol-gel transition*)—must be overcome. These gels are reversible because the connections between the macromolecules of the gelling agent are temporary. They are also called *physical gels*, for the bonds—hydrogen or electrostatic bonds, or bridges between the hydrophobic portions of the macromolecules (as with cellulose or nonelectrically charged polysaccharides)—are physical bonds. These weak bonds permit the gel to dissociate as a result of the effect of weak changes (change in temperature, addition of ions) and to reconstitute when the pieces are reconnected. The absorption of a solvent produces unlimited swelling of the gel when the temporary bonds are too weak or too few.

In contrast, when they are too strong or too numerous, they lead to gel collapse, like a polymer in the presence of a bad solvent, and the gelling agent and the liquid "divorce." Wait around long enough and the number of bonds may increase spontaneously; gels are sensitive to the ravages of time. This phenomenon of expelling solvent—called *syneresis*—is the constant dread of manufacturers of gel products. Like emulsions, gels are unstable or *metastable* systems whose character is specifically determined by the linking behavior of the macromolecules of the gelling agent. Different states are possible, and when the gel is in one of these states, it is susceptible to evolving into another one—better adapted to the circumstances—depending on the temperature or a mechanical constraint that changes the distribution of the components.

Other gels make strong, permanent bonds—chemical gels, like cooked egg white, which is essentially a network of proteins that trap water (90 percent of the white). For superabsorbent products, like baby diapers, the cellulose fibers in the pants contain little balls, each made up of a network of cross-linked hydrophilic polymer chains. The network swells and captures the liquid, stabilizing once it has reached equilibrium, and surrenders no fluid (or only a little) in response to mechanical pressure. Thus, it is not a sponge, and the baby can sit down. The gelling agent–polymer can hold a mass of water several hundred times its own mass.[4]

It is also possible to have gels with no bonds between their macromolecules; such macromolecules need only be sufficiently tangled for their movements to be restricted, which prevents liquid from escaping. For example, the addition of 0.1 percent of xanthan gum is enough to keep water from flowing. Excreted by the bacterium *Xanthomonas campestris,* this commercial polymer, in the form of a cream-white powder, outsells all other microbial polysaccharides. The industrial applications made possible by its exceptional properties are many: oil recovery, cement manufacture, paint formulations, and especially foodstuffs.

Given a liquid that proves a good solvent for macromolecules of the gelling agent, why does the gel remain swollen, and why does the liquid not escape the fairly benign net in which it is caught? One reason only: osmotic pressure. This pressure derives from the attraction between the liquid molecules and the surfaces of the net and vice versa—a play of reciprocal and interactive influences . . . a veritable osmosis. Crude dispersions—consisting of clusters of gelling agent or of aggregated particles—have very weak or no osmotic pressure and consequently retain liquid only poorly. The osmotic pressure of a gel is thus linked to the dispersion state of the gelling agent. This pressure tends to resist the forces of attraction between dispersed objects. Depending on the balance between the two tendencies—repulsion and attraction—aggregation is inhibited or promoted, and the gel remains swollen or collapses. A very small quantity of gelling agent can cover an enormous surface—the selling point, as it happens, of the finely divided matter we call a colloid. For this reason, the low-profile presence of this additive in a liquid can create a material with totally new and specific properties relative to those of the pure liquid, for osmotic pressure depends on the existence of this surface. This direct link between surface and retention of liquid gives rise to the sensitivity of the liquid material to a very small amount of additive.

When does gelling start? Asking this question means tackling the issue of *percolation*. The hot water in your espresso percolated through ground coffee on its way to your cup. A fire declared in several areas of a pine forest percolates when all of the hot spots connect. Connectivity is also a fundamental element of social life; percolation

> ### Yes to Blending!
> Salt preserves meat and fish, just as sugar does fruit, because it destroys bacteria by osmosis. A bacterium is an aqueous solution bound by a thin membrane. Water is drawn out of the cell by the powder, and the dehydrated bacterium dies.
>
> For the same reason, pure water is dangerous for our cells, which contain salty fluid. That is why cleaning your eyes with water instead of physiological serum is discouraged. The water goes into the cells of the cornea and can cause swelling. The same is true for soft contact lenses, which are hydrogels kept moist with saline solution.
>
> Salting a steak draws the blood out. The best situation is one in which the water molecules are mixed equally with all of the others, those caught in the steak fibers and the powder on the outside; the steak is in osmotic equilibrium.
>
> In one sense, osmosis is the principle behind the right to blend.

makes it possible to treat a material—sensitive, I might add—in the same way that terrorists infiltrate a country through the passivity of its citizens—through people not directly involved in the acts but who do not oppose them or who do not communicate what they know although they could. To make an analogy, each sympathizer is a macromolecule of the gelling agent. The group of terrorists can move freely throughout the targeted territory as soon as its cause is passively supported by 13 percent of the population.[5] That is an extremely small number and is even counterintuitive, which is why it is so hard for the authorities to stem the menace in the countries where it has become institutionalized. Approaching the problem as one of percolation provides a way to do so, however, by suggesting that measures be developed that are complementary to those based on destroying a cell since a new group has only to reorganize for terrorism to strike again. Percolation is a sudden phenomenon, a transition that produces a radical change from one state to another.

For a solution on its way to becoming a gel, how many junctions are needed to percolate? If the molecules of the gelling agent are weakly interpenetrated, the number of junctions required for percolation is equal to the number of macromolecules—a lot! Moreover, increasing the number of junctions is not particularly helpful: It can lead to gels with limited swelling or that spontaneously collapse. In contrast, if the macromolecules are strongly interpenetrating, the solution can percolate with a much smaller number of junctions, many fewer than there are molecules. Gelling proceeds poorly when the macromolecules are bushy or dense objects because the concentrations required are too high to be practical. A good rule of thumb: Gelling is much more effective and more homogeneous when the macromolecules are elongated, strongly interpenetrating, and weakly cross-linked. Sensitive cooks, take heed!

STABILIZE

The fact is that everything we eat is a dispersed medium. Foods are complex systems with multiple constituents and phases. The dispersed entities are sometimes perfectly spherical—gas bubbles, droplets—and sometimes only roughly so—fat globules, proteins, starch granules. Their size varies from nanometers (surfactant molecules) to micrometers (droplets) to millimeters (bubbles). The most common dispersed phases are liquid water (or an aqueous solution), liquid oil (or partially crystallized fat), and air (or carbon dioxide). The majority of food colloids are emulsions or foams rather than dispersions of solid objects in a liquid. There are emulsions of oil in water—milk, cream, mayonnaise, cake batter, many sauces—and emulsions of water in oil—margarine, Nutella, vinaigrette. Ice cream beats all other food colloids for complexity. It is simultaneously a foam, an emulsion, a dispersion, and a gel and has a state that is partially gas, liquid, crystal, and vitreous. It contains droplets of fat, air bubbles, and ice crystals, all dispersed in a continuous aqueous phase and aggregated to form a frozen, semisolid, aerated matrix with a light, creamy texture!

Stability is a critical feature of these heterogeneous mixes inasmuch as it preserves their character, mechanical structure, and shelf life. Equally important and a focus of intense research is a controlled instability, once in the mouth, that gives a food its sensory quality. The interaction of an emulsion with saliva and the impact of the mechanical constraints associated with manducation (processes that precede digestion, including chewing) are hot topics for researchers in food science. The concept of the stability of a colloid is primarily related to time. Gravy (an emulsion) has a short shelf life, which one learns to deal with, whereas a cream liqueur must last for years. Stability implies no perceptible change in the structure of a material. Stabilizing is also related to texturing, which is another important objective of a cook, pastry chef, or baker. Whether a pastry dough is puffed, shortcrust, or crumbly actually depends more on its mechanical preparation than on its composition.

What misfortunes await an unstable emulsion made of barely miscible liquids? First, creaming, which occurs when oil droplets form a concentrated layer at the top of an emulsion of oil in water. Then there is flocculation, which is generally reversible: The particles associate but remain separate, which is different from coagulation because the forces between coagulated particles are stronger. Flocculation often precedes creaming and coalescence, which is the irreversible fusion of one or several droplets of an emulsion into a single, larger one. This latter form of destabilization is the most serious. A consumer who detects it will put a product back on the store shelf. The problem of the stability of emulsions and foams is an active area of exploration because of its complexity: There is no single mechanism at work. Scientists are seeking common behaviors in apparently very different systems.

What can be done to avoid or at least to delay the appearance of an instability in a food colloid? First, one has to distinguish between the tendency of an additive to emulsify and its capacity to preserve an emulsion over time. Proteins basically act at interfaces (oil-water for mayonnaise, air-liquid for champagne); they are frequently both emulsifiers and stabilizers. On the other hand, carrageenan and xanthan are only stabilizers, for they have no surfactant properties (the molecules are not amphiphilic). They act by forming a gel or by modifying the viscosity of the aqueous phase. Increasing the viscosity of a food colloid stabilizes it. It is an alternative to stabilizing the interfaces with amphiphiles. Between one and four grams of xanthan per liter of pulp-rich fruit juice will keep the strands in suspension. The formation of a very viscous phase will also keep the honeycomb structure of a mousse from draining too quickly and breaking. To make a homogeneous raisin cake, roll the raisins in flour before adding them. A light dusting of flour is enough to alter their surface roughness and keep them from drifting as the cake rises; the force of gravity will have a harder time attracting them to the bottom of the cake. Of course, this precaution is unnecessary if the viscosity of the batter is sufficient from the very start.

COAT

Coating an ingredient forms a protective barrier against water, oil, or gas. According to chefs, it is a way of *sealing in* aromas and flavors by limiting the influence of other ingredients. Coating stabilizes the form of an ingredient and protects its integrity at high temperature: Drying removes water, and as the temperature decreases, molecular reactions and crystallization are favored, which leads to undesirable coagulation and phase separation. Coating may also add color by contributing a specific ingredient that will brown, like the fine film of egg yolk that a baker brushes on a *viennoiserie* before putting it in the oven.

There are all sorts of natural systems for which a liquid, a gel, an emulsion, or a foam are demarcated by a surface envelope. For fish or meat, it is a network of fibers. The juice of fish eggs or of currants is released after the envelope is punctured. One of the celebrated practices of molecular cooking is *spherification,* which consists of forming pearls that are solid on the surface and liquid inside. Dissolve at most 1 percent sodium alginate in juice or sauce. Add drops of this solution to water containing from 1 to 4 percent calcium, for the alginate will gel in the presence of a salt such as calcium salt. A soft skin will immediately form around the drops. Rinse. The result will be liquid marbles of melon juice, red wine sauce, or raspberry puree. Beware of juice that naturally contains calcium or hard tap water because they will make your marbles hard on the inside. For the same reason, forget about making liquid marbles from milk. Very acid juices and strong alcohol are similarly ill suited to the technique. Liquid marbles are unstable over time—they will shrink and release the liquid—but you will hardly wait that long to succumb to the pleasure of having the plump, tasty pearls pop in your mouth. Dosage, time, and the proper amount of salt are the essential parameters of success in this example of designer cuisine.

THICKEN

Thickening is what cooks call *binding*. In small quantities, gums (xanthan, guar, locust bean) are excellent thickeners. The sauce chef coagulates the proteins: egg, in the case of béarnaise sauce, or blood, for stew. The chef may also use starch: wheat flour (which is 70 to 80 percent starch), rice, potato, corn, tapioca. Heated in liquid, starch swells on absorbing the liquid and gels. Gelling thus often serves to thicken a preparation. However, you can also thicken without gelling; that is what the sauce chef does in adding fat to a liquid to make an emulsion (*beurre blanc* sauce). In any case, the goal is to obtain a preparation that is neither too liquid, which would be a juice, nor too solid, which would be a puree.

> If you cannot manage a bit of sorcery, you should stay out of the kitchen.
>
> —Colette, *Prisons et Paradis*, 1932

ADAPTATION

RESPONDING TO THE ENVIRONMENT

Nothing endures but change.
—Heraclitus of Ephesus (6th c. BC)

Thriving through change: How to adapt to and deal with new situations
Day 1
 • Sizing up new situations
 • Learning to recognize ways of adapting and unexpressed needs
 • Dealing with challenges
Day 2
 • Building self-confidence
 • How to be in a winning position to take the next step forward
Cost of registration: $1,500, excluding tax

—Seen in a continuing education catalog.

CHAPTER 11

A CELL, THOUGH NOT A PRISON

TWO OR THREE THINGS WE KNOW ABOUT IT

From the most majestic redwood to a unicellular bacterium, the cell is the smallest unit of living matter. No life is possible without it. The cell contains everything needed to survive in a perpetually changing environment. Although most cells are only a few micrometers in size, some, like neurons, can be as long as a meter. The cell is bounded by an extremely fine envelope that is seven to eight nanometers thick—the plasmic membrane. It encloses a solution, the cytoplasm, which contains thousands of different molecules. The membrane consists of a double layer of amphiphilic molecules, mostly phospholipids with a hydrophilic head connected to a hydrophobic tail made of two chains of fatty acids. The chains face each other, whereas the heads are exposed to water both inside and outside the cell. The cellular membrane is a natural liquid crystal with a layered structure.

The cell needs its membrane to protect its vital molecules (DNA and RNA) from the outside environment, maintain the balance of the cytoplasm, communicate with other cells, and provide a physical place of exchange (supply of nutrients and evacuation of waste). The cell membrane is thus far from being an impenetrable wall, like that of a prison. It is, however, merciless in selecting the entities that are capable of either penetrating the cell or leaving it.

CELLULAR TRAFFIC

Our cells swim in a liquid derived from blood that contains thousands of ingredients: amino acids, sugars, fatty acids, vitamins, mineral salts, ions, hormones, carbon dioxide, oxygen, and waste. Each cell must extract from

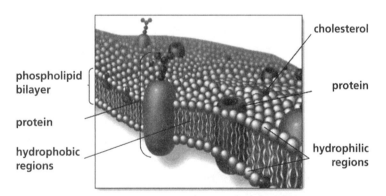

The cellular membrane: a double layer of amphiphilic molecules (mostly phospholipids) encapsulating different proteins with diverse functions, cholesterol molecules, glycolipids, and carbohydrates (not shown here).

this milieu the substances it needs at any given moment. This trafficking of ingredients may occur either passively or actively.

In passive mechanisms, molecules pass through the membrane without the cell having to exert any energy. This is the case with molecules that are small enough to move through the membrane simply by squeezing between the molecules of the bilayer to balance their concentration on both sides of the membrane. The molecules' energy of motion (kinetic energy) drives this diffusion. Small ions or small electrically charged molecules are powered not just by chemical forces—linked to concentration differences of various chemical species—but also by electrical forces; different electrical charges inside and outside the membrane either facilitate chemical diffusion or oppose it.

In contrast, when molecules are too large or when an ingredient is made to move counter to its natural tendency, active mechanisms take over. In exchange, the cell calls on fuel—adenosine triphosphate (ATP). Some of the energy contained in the food we eat is stored in our body in the form of chemical bonds in this little molecule. The storage is temporary. Whenever cellular processes require energy, the ATP bonds are broken, and energy is released. Among these active mechanisms, one is extremely important for animal cells and is also a very good illustration of the necessary and beneficial sensitivity of the plasmic membrane: the sodium-potassium ion pump. This mechanism is critical to excitable cells such as nerve, muscle, and hormonal cells. More than half of the energy expended by a neuron goes into operating sodium-potassium pumps.

THE MEMBRANE HAS NO LACK OF POTENTIAL

When a cell is at rest, the concentration of potassium ions (K⁺) is ten to twenty times higher inside the cell than outside it. The inverse is true for sodium ions (Na⁺). The natural tendency of K⁺ is to migrate to the outside of the cell to balance its concentration in the intra- and extracellular environments and to seek osmotic equilibrium. The tendency of Na⁺ is the opposite—but for the same reasons. Now, it so happens that at a state of rest, the membrane is permeable to K⁺ but nearly impermeable to Na⁺. The diffusion of

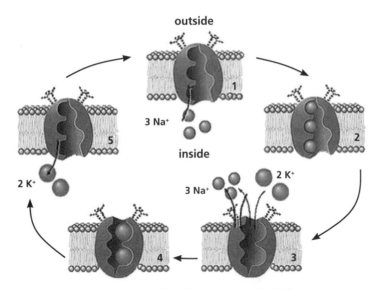

The sodium-potassium pump: sodium (Na⁺) and potassium (K⁺) ions pass through protein transport channels on either side of the cell wall following precise instructions. Events (1) and (2): three Na⁺ ions from inside the cell enter the channel, which recognizes them. (3) These are released to the exterior. Two K⁺ ions arrive and (4) enter the channel, which recognizes them. (5) These ions are released inside the cell. This cycle corresponds to a pump stroke.

the two types of ions across the membrane is unequal, which produces a deficit of positive ions inside. Thus, the membrane does not have the same electrical charge on either side. The result is a difference in electrical potential called *resting potential*. All it takes is a surplus of around two ions in one hundred thousand to evoke it. The resting potential is characteristic of every living cell membrane. An animal cell membrane has a resting potential that ranges from −20 to −200 thousandths of a volt (mV) depending on the organism and the cell type. The negative sign indicates that the internal wall is negatively charged with respect to the external wall. The electrical voltage matters only at the level of the membrane: If you add all of the positive and negative charges contained in the cytoplasm, you will find that the interior of the cell is electrically neutral. The same goes for the extracellular environment.

During activity, excitable cells are capable of producing one or several transitory and sudden variations in resting potential. The channels opened by this change in potential are carried by proteins that cross the entire thickness of the membrane and constitute a pore. Depending on the value of the potential, these channels are in an open state—ions pass through the pore—or a closed state. The sodium-potassium pump continually maintains the imbalanced distribution of Na^+ and K^+ ions on either side of the membrane by linking the transport of one to the other. Each stroke of the pump causes three Na^+ ions to leave the cell and two K^+ ions to enter. Because the membrane is more permeable to K^+, the ionic imbalance and the membrane potential are conserved. If Na^+ ions begin flooding into the cell, the pump activates to push them back out and to maintain the essential disequilibrium. If sodium were not continually exported, it would accumulate in the cytoplasm, and osmosis would take over, drawing water into the cell . . . until it burst.

The essential modification of the membrane's resting potential corresponds to a phase known as *depolarization,* during which the (resting) potential becomes *action potential* and reaches its peak value at its apogee. During this time, three Na⁺ ions leave the cell. Action potential is the basis of nerve conduction. This modification is local. Nerve

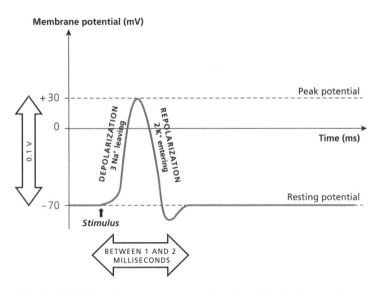

Evolution of cellular membrane potential as a function of time during a stroke of a sodium-potassium pump. Following a stimulus, the membrane undergoes depolarization, during which its polarity reverses and the potential goes from a resting value to one of action, which reaches a peak value—all simply to gain a tenth of a volt of amplitude. During this phase, three Na⁺ ions pass from inside to outside the cell. Depolarization is succeeded by repolarization, during which two K⁺ ions make the opposite trip. The entire cycle takes only one or two milliseconds.

conduction is the displacement of action potential along the neuronal membrane. The polarity of the membrane reverses in a mere one or two thousandths of a second. The potential goes from −70 mV (the average resting potential for more than half of all cells) to +30 mV, the average peak action potential—in other words, an amplitude of barely a tenth of a volt. *Repolarization* follows depolarization: Once the positive peak value has been reached, the action potential diminishes until it regains the resting potential. During this phase, two K^+ ions enter the cell.

The cellular membrane is an autoadaptive, intelligent sensitive material because of the organization and selectivity of the processes that take place there in close association with the changing environment . . .

CHAPTER 12

PUTTING DRUG DELIVERY ON

CONTROLLED RELEASE

No wonder people dream of making structures similar to that of the cell to use as soft capsules for transporting drugs through the blood for delivery at just the right time and place!

MEETING UP, BUT WHERE AND WHEN?

Whether by pill or by injection, drug therapy comes down to achieving both optimal action at a certain time after delivering the agent and subsequent subsidence. The cycle repeats with the next dose. However, the doctor wants the dose of the active molecule in the blood to remain below a critical threshold, at which there is a risk of drug-induced toxicity or intolerance, the formation of undesirable products resulting from the drug breaking down, or some other

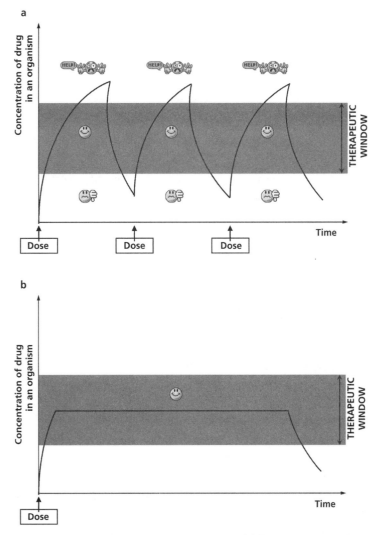

Evolution of a drug dosage in an organism over time. (a) The classical scenario: the dose increases and exceeds a threshold beyond which side effects may occur, then decreases to another threshold, below which too little of the agent is present to be effective. This sequence repeats with the next dose. (b) The aim of controlled release: a single dose confined between two warning thresholds at a fairly constant level for a much longer period of activity than that of classical therapy.

constraint (such as patient discomfort or surgery to insert or remove the device that delivers the drug). By the same token, the doctor wants the drug dosage to remain above a minimal value, for too little will not have the desired effect. The two thresholds define the *therapeutic window*, and the aim of a controlled-release drug delivery system is to satisfy both objectives. The goal might be to give a drug in a single, much-reduced dose. Given in constant quantity over a longer period, the dose would not fluctuate. After administration, the moment of release of the drug might be controlled by so-called *chronotherapy*. For example, cardiorespiratory function fluctuates throughout the course of a day. Thus, for asthmatic patients, the objective would be to have the release of the active ingredient coincide with critical periods.

In addition to the issues of time and duration, the delivery site, referred to as drug *vectorization*, also presents a problem. In cancer the diseased cells are localized. However, chemotherapy commonly administers drugs to the entire body, which maximizes side effects. The aim of controlled drug delivery is to direct anticancer molecules only to malignant cells.

The active ingredient must be encapsulated and protected so that the carrier-cargo assembly is delivered to the site, the cargo unloaded, and the carrier then made to disappear. Indeed, the carrier: which one, exactly?

ENCAPSULATION-DECAPSULATION: LIPOSOMES

Liposomes—from the Greek *lipos*, fat, and *sôma*, body— are vesicles, little hollow spheres that encapsulate an aqueous solution whose walls are a double layer of amphiphilic molecules; the lipophilic tails are together, and the hydrophilic heads are in contact with the aqueous medium.

As early as 1911, Otto Lehmann (the physicist we met earlier in our discussion of the discovery of liquid crystals) had unwittingly made liposomes while studying blends of lecithin, cholesterol, and water—which also brings to mind our accounts of mayonnaise and dissolving fats. The size of liposomes varies from a few tens of nanometers to a few tens of micrometers. Their membrane is only a few nanometers thick. The molecules are able to move laterally in the plane of the membrane while in general maintaining their orientation. Like its cellular homologue, the membrane is a liquid crystal. The role of liposomes is to transport constituents to be incorporated into the cytoplasm by interacting with the cellular membrane according to three possible scenarios. First, adhesion: A liposome lands on the membrane,

Cross-sectional view of a liposome. The membrane is a bilayer of amphiphilic molecules that encapsulates an aqueous solvent containing the drug.

establishes direct contact with it by anchoring itself to it, and releases its contents by diffusing them through the double wall—that of both the liposome and the cell. Permeability is achieved through a defect at the site of contact. A second scenario: fusion. The liposome membrane fuses directly to the cell membrane. The proteins of the liposomal membrane diffuse into the cell, and the cargo is spilled directly into the cytoplasm. The last scenario: engulfment or *phagocytosis*, a much more elaborate operation and one that is often described as the most frequent process occurring in vivo. The cellular membrane engulfs the liposome, allows it to penetrate, and then self-heals once it is inside. It is possible to promote one of these three basic mechanisms by modifying the composition of the bilayer. In chemotherapy,

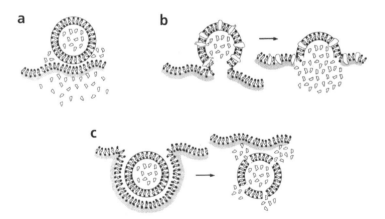

Three mechanisms of liposomal drug delivery in a cell. (a) Adhesion: the liposome adheres to the cellular membrane, and its contents spread through the two walls. (b) Fusion: the membranes open and then fuse, accompanied by progressive release of the contents of the liposome, such as protein cargo (shown as having been an integral part of the membrane). (c) Engulfment: the entire liposome is engulfed by the membrane and disintegrates inside the cytoplasm.

anticancer agents are transported to the tumor target. In gene therapy, the genes are led to the nucleus of the cell to be expressed as a way of delivering nucleic acids by providing it with a missing gene or, conversely, to inhibit the expression of a pathological protein (viral or cancerous).

The selection of transfer sites is made by developing liposomes that are sensitive to temperature, pH, or the character of the membrane proteins of the target cell.

For a liposome to release its cargo at a certain temperature, the properties of membrane phase change must be exploited. Stable at 37°C, it becomes softer at 40°C owing to the alteration of the electrostatic interactions between molecules. In the case of the local temperature of a tumor or a part of an organism that has been heated externally, fractures appear in the wall of the liposome, which breaks, sending its contents, including the precious active ingredient, toward the tumor.

The sensitivity of pH (a measure of the acidity or basicity of a solution) is put to good use in the case of phagocytosis. After engulfment by the cell, liposome envelopes carrying genetic material are destroyed in the cytoplasm, where the pH is lower than outside—which is exactly what is desired in terms of delivery site. But how to do this? The trick is to choose amphiphilic molecules whose heads bear an electrical charge that is conserved in neutral pH—the membrane remains stable thanks to the electrostatic repulsion between the heads—but is lost in the cytoplasm, at a weaker pH, which destabilizes the membrane and leads to its disintegration.

Selective disintegration of liposomes on the approach to the specific target relies on a similar strategy. The membrane is stabilized by including proteins chosen as antibodies

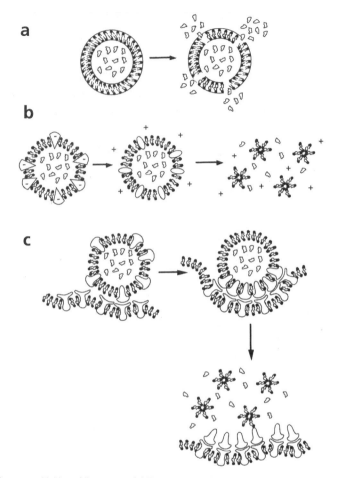

Three sensitivities of liposomes: (a) Temperature. The membrane softens with an increase in temperature. The content exploits this weakness and escapes through structural defects in the membrane. (b) pH. When the acidic or basic nature of the environment outside the liposome changes, the stability and structure of the membrane change, too. The content is released, and the amphiphilic molecules of the liposomal membrane assemble differently (inverse micelles), adapting to the new conditions. (c) Composition of the cellular membrane. The liposomal membrane comprises proteins that complement those of the cellular membrane. When the liposome arrives in the vicinity of the cell, its proteins are attracted by those of the cell. They move toward the liposomal membrane and pair with their cousins, a first step in leaving home. The liposomal membrane loses its integrity. Here, too, the constituents assemble into inverse micelles.

of an antigen specifically designed as a target. The liposomes remain stable if the antibodies do not encounter the receptor. When there is an encounter, the antibodies reassemble and attach to the corresponding site, like a key in a lock, because the antigen-antibody complementarity is stronger than the interaction between the constituents of the membrane. The membrane is destabilized, and the liposome releases its cargo.

SWELLING-DESWELLING: EXCITABLE POLYMERS

Research on controlled delivery systems focuses particularly on hydrogels, which consist of polymer chains that trap between 60 and 90 percent of water. What determines delivery, depending on the polymer, is temperature, pH, the concentration of ions in the environment, the chemical composition of the arrival site of the system, or ultrasound vibrations.

One polymer that is soluble in water at low temperatures but precipitates when the solution is heated is N-isopropylacrylamide (NIPAAm). This behavior is surprising at

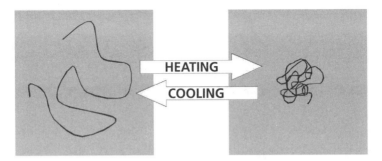

A molecule of NIPAAm polymer in water. Left: at low temperatures, the chain relaxes in the solvent. Right: at high temperatures, it forms a ball to minimize contact with the water.

first because we are used to seeing substances become more soluble in hot water (you can melt more sugar in hot coffee than in cold coffee). The explanation lies in the chemical nature of the polymer molecule, which has segments of different kinds: one soluble in water and one not. In solution, the first interacts willingly with water and favors an unfolded polymer molecule, whereas the second retracts to avoid contact. The unfolded state of NIPAAm is stable at low temperatures because of bonds that form between the soluble segments and water molecules. This relative immobilization of the water molecules is the price of stability. We say that their *entropy,* a measure of the degree of disorder of a system, is reduced. However, if entropy decreases, the quantity of energy the system loses must be compensated for, according to the first law of thermodynamics, which states that energy is conserved during transformation.[1] Actually, heat is produced; it results from the formation of bonds between the water-soluble segments and water molecules. Yet, and this is the essence of the explanation, a reduction of entropy and the release of heat will not achieve perfect balance beyond a certain temperature. The polymer molecule then expels most of the water molecules around it, and the presence of water-insoluble segments causes the polymer molecule to ball up. This behavior is reversible: If the temperature is lowered again, the inverse transition occurs: balled chain to extended chain. So much for the isolated behavior of a NIPAAm molecule in water.

Now let us connect a group of these chains to form a hydrogel in water. By changing the number of links and the quantity of solvent, the consistency of the gel will vary from jelly (fewer links, more solvent) to rubber (more links, less solvent). When the temperature reaches a critical point, the chains form a ball, and the solvent is expelled from the gel—like squeezing a sponge. If the solvent contains an active ingredient, it is released at the site.

The critical temperature is regulated by the relative number of soluble and insoluble segments in the polymer molecule. For certain ratios, the swelling-deswelling transition of the gel is very sensitive to a very specific temperature: The temperature need vary by only one or two degrees around the critical temperature for an abrupt change in volume to occur. The fabricator thus has the possibility of targeting the delivery temperature. The time it takes the gel to change volume depends on its size.

For other NIPAAm gels, a change in pH drives the transition. The active ingredient will not be delivered at a weak pH, in the stomach, but where it needs to act, in the small intestine, at a higher pH. Yet other gels can swell or deswell in response to an electric field, which suggests the possibility of an electrically controllable valve or an artificial muscle. Systems based on NIPAAm are among the best available

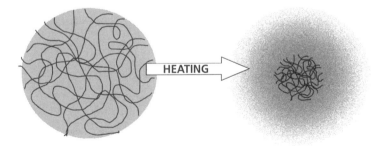

HEATING

Behavior of a ball of cross-linked NIPAAm in response to water. Left: at low temperatures the ball swells because the polymer likes water. At right, the gel is deflated: the polymer collapses because the temperature has risen beyond the threshold at which the polymer separates from the water.

candidates for controlled drug delivery. From a practical point of view, there is considerable interest in finding materials whose change in critical volume is compatible with physiological pH (7.4) and body temperature (37°C). Developing a smart polymer will require a two-part molecule whose responses are the diametric opposites of a given stimulus.

Researchers have fabricated a polymer that encapsulates tiny magnetic particles for use as a drug carrier.[2] When the material is subjected to an oscillating magnetic field, the particles jump, opening pathways in the tangled polymer network that serve as escape routes for the drug. When the field is cut, the release stops or markedly decreases because the pores close. The magnetic field would be delivered by a device the size of a wristwatch that the patient can activate

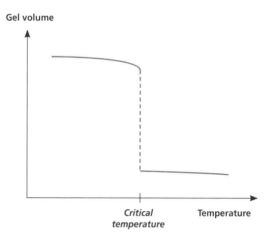

An example of the high temperature sensitivity of NIPAAm. A slight increase in temperature can cause the sudden and spectacular collapse of the gel. This leads the gel to expel its solvent, doped with active agent, into its environment at a specified temperature.

at will. This is particularly important in treating diabetes, where the amount of insulin delivered is based strictly on blood glucose levels over time. Another approach to the same problem is to use *magnetoliposomes,* that is, liposomes made of a colloidal suspension of nanometer-sized magnetic particles suspended in a liquid. An alternating magnetic field converts magnetic energy into heat. The magnetoliposomes are drawn into the tumor by external magnets and, when heated, release the active ingredient. This magnetic hyperthermia is one promising way of targeting cancerous tumors.[3]

Another example of stimulus is vibrations. An ultrasound beam is directed at the tumor to be treated. Micelles at the tumor release their contents either because they have been weakened or destroyed by mechanical shock or because the cellular membrane has been made more permeable—or for both reasons.

BEST OF LUCK!

There are many candidate drugs for distribution by control of their site and time of action: chemotherapeutic agents, immunosuppressors, anti-inflammatories, antibiotics, hormones, steroids, anesthetics, vaccines, and cosmetics. The major problem of some carriers is that they are too rapidly removed from the blood by metabolism. Consequently, they become concentrated in the bile, liver (whose mission is precisely to destroy foreign substances), or other places in the cell to which they are toxic. Another restriction of carriers is degradation by tearing rather than dissolving. Research is investigating hydrogel formulations in which the surface of the polymer envelope erodes slowly—like a piece of candy that you suck on—as opposed to fracturing.

Thus, solutions continue to be needed to prolong the residence time of doped carriers in the blood prior to their arrival at the target. Once there, the vehicles need to decompose and allow the organism to digest the remains naturally. Controlled delivery that works closely with physiological changes is still a challenge. The stability of carriers must be improved (avoiding degradation of liposomes by enzymes), methods of fabrication need to be simplified, and their sites of action must be more precisely targeted. Liposomes are being successfully employed to transport anticancer treatments and remain good candidates for gene therapy. Molecular recognition is an objective of primary importance; much promising work is already well under way, but there is still some distance to go.

A cell measures only a few micrometers in size, which means that chemical reactions are taking place in compartments that hold a few nanoliters. Circulation through the cellular membrane is carried out by pores of only a few nanometers. The aggressors of cells (bacteria, viruses) have a size intermediate between a micrometer and a nanometer. That is why the conception of controlled drug-delivery systems is an essential subject of research in *nanotechnology*.

The idea of manipulating the smallest constituents of matter—atoms and molecules—which is the original objective of nanotechnology, goes way back in the scientific imagination. Only relatively recently, however, has it begun to attract grants from funding agencies and the attention of the media. The latter sometimes seize on it to publicize or to arouse all sorts of fears about the subject. As citizens, we want to be able to question the purpose of scientific research, and our desire is legitimate. Nonetheless, our thinking is often influenced by the confusion made, in

the information we are offered, between scientific activity—
and the results it produces—and the use of these results. We
must address concerns about nanotechnology by soberly
debating its objectives. Should the cautionary principle be
applied? Unfortunately, it is often confusingly mixed with
the principle of inaction, which risks threatening the advance of knowledge. If the intention is to achieve a morato-
rium, and taking as an example the problem of the controlled release of nanoencapsulated drugs—are we really
willing to forego the sophisticated tools we expect from
progress in nanotechnology?

CHAPTER 13

PERPETUAL SENSITIVITY:

GRANULAR MATTER

AGE-OLD MATTER, FRESH TOPIC

Granular matter merits a special stop on our itinerary, for its behavior is not determined at the molecular level but at that of objects visible to the naked eye—or nearly. The size of these grains—between a micrometer and a centimeter— makes the influence of thermal agitation negligible and any comparison with the behavior of molecules in fluid irrelevant. Yet sand flows between our fingers. Like a liquid, it takes the shape of its container, and, seemingly like a solid, it forms an immobile pile on a table. Landslides and avalanches attest to the double life of granular material—liquid, yet solid.

If the material most used by humans in terms of quantity is water, granular matter probably runs a close second.

However, the answers to the scientific questions it raises are still in their infancy. We have no unified vision of the subject. For a long time, mirroring the trajectory of liquid crystals, research in the field went along without funding or encouragement. The situation changed somewhat once decision makers understood that very concrete technological problems—for example, in the construction, agrifood, chemical, and pharmaceutical industries—rely on our understanding the complex behavior of granular matter. Decision makers are more amenable to these kinds of arguments than to helping support the cultural role of research over the long term. Yet the real promise of the field—the applications of the distant future—depends on having a fundamental understanding of a problem.

Grains of different sizes, shapes, and surface roughness will not mix when shaken. When flowing over a slope, grains tend to separate even if the starting mixture is homogeneous: The largest rocks are found at the base of scree slopes. Because pharmacology mixes many different powders and grains and developing a drug necessarily implies combinations, understanding the physics of the phenomenon is imperative. Thanks to the power of computers, we can formulate equations of motion that predict the evolution of dunes. The threat to countries encroached on by deserts, such as Mauritania, is driving experiments both in the field and in the laboratory. One objective is to halt the advance of *barchans,* sand dunes in the form of a crescent that rise to more than thirty meters in height as they propagate, obliterating rare vegetation and obstructing access roads and inhabited areas.

COMPLEX MATTER, BUT WHY?

The surge of interest in work on granular matter is both to understand the laws governing the interaction between two grains and to describe their collective behavior: Whatever the size or nature of the grains, different piles must share certain characteristics. It is still very difficult to describe the contact forces that result from shocks and friction and to know how energy is dissipated in the course of these events. Granular material is yet another example of sensitive matter giving rise to new research subjects in physics that do not require complex instrumentation or theories.

In a cubic millimeter of liquid, one can blithely ignore what each individual molecule is doing and concentrate instead on their collective behavior, for there are billions and billions of molecules. Such systems are described in terms of average parameters: This is the case with temperature and pressure. A cubic millimeter of sand, on the other hand, contains a small number of individual entities that are easily countable using a magnifying glass and tweezers, and one cannot avoid describing their individual behavior. How does a grain of sand move? Bump another grain? Rub against it? Suddenly the behavior of a granular material depends heavily on the personality of the individuals that compose it: Thus, coming up with generic behavior is a complicated enterprise. The interactions between grains are difficult to model because the grains have different sizes, shapes, and surface roughness. Moreover, a grain of sand is not a grain of wheat or a polymer granule; a grain of ice is highly complex because the water can assume different states.

In research activities on granular matter, the initial state must be defined very carefully. One cannot simply write the following in an experimental section of a scientific paper: "We fill a glass tube with sand." Do you fill it from a jet of sand along the axis of the tube? Or do you sprinkle the sand over the whole tube opening? Do you shake the tube after filling it? All of these questions are more than mere details for the evolution of the system and its understanding.

THE PARADOX OF THE MISSING MASS

. . . paradoxes, this smoke-without-fire of the intelligence.

—Frédéric Vitoux, Sérénissime, 1990

The flow of a granular material is subject to sudden and catastrophic jams. In the case of an hourglass, a funnel, or a silo, the operator should lose no time in attempting to increase the size of the conduits and the receptacles. In addition, it will be necessary, from time to time, to give them a few thumps to reestablish the flow. The source of the blockage is elsewhere: It is discreetly hidden in the mass of material as the arching of a portion of the grains—like the stones of a church vault. If one were to continually fill a silo with water on a sufficiently large scale, at every instant the scale would register the exact quantity poured in. What happens when one pours in grains in a constant and continuous flow? At first, the mass weight equals that of the grains poured. Intuition is validated. However, the measurement very quickly reaches a maximum value (as soon as the height of the pile approaches the diameter of the silo). Where has the mass disappeared to? The grains have formed myriad vaults within the mass of matter. Their weight is not transmitted vertically toward the bottom but along the arches and toward the lateral walls of the container, which partially support it. The weight measured by the scale reflects only the weight toward the bottom. The weighed mass is apparently that of the grains whose weight was not subtracted from the whole by the walls . . . which are, one hopes, especially solid. Furthermore, the greater the friction on the walls, the more likely vaulting becomes. Any additional quantity poured into the top of the silo will be entirely supported by the walls. If a grain silo remains unused for a certain period of time, the grains will stick together, and the vaults will solidify. They create cavities in which fermentation gas accumulates: As pressure increases, the silo may explode. Another source of explosion is the electrical charge picked up by the grains during collisions, such as those with the walls of the silo.

KEY TO THE VAULT OF SENSITIVITY

Despite its appearance of solidity, stability, and permanence, a pile of granular matter is very sensitive to the least perturbation. A tiny shock is enough to alter the way in which the grains are arranged and, in turn, the structure of the vaults, which starts the pile flowing again. The vault network may also change in response to a slight variation in temperature. In an hourglass, a small change in temperature between the lower bulb (held in the hand) and the upper bulb (exposed to the air) suffices to stabilize a vault for a few moments. The propagation of a vault depends on the grains that support it. If one of them slides just a bit or modifies its area of contact with a neighbor by presenting another area of its surface, the local vault network is also susceptible to change. An outside observer will perceive nothing—the pile remains generally immobile—but measuring the apparent mass over time will show that immobility is deceptive. Small events can trigger large effects.

Thus, a solid-liquid transition occurs when the arches of a grain silo are subject to a small shock. The same transition is evident in an avalanche, but what causes it is different. An avalanche is unleashed when the slope of granular material exceeds a certain limit, the *maximum angle of stability*. The movement of just a few grains is enough to trigger the avalanche: One grain tumbles down the slope, hits another, which starts rolling in turn, and so forth. At the conclusion of the event, the slope will have diminished by a few degrees to once again attain its *angle of repose*. Barchans (dunes) can advance up to sixty meters per year by a succession of avalanches: collapse, rest, collapse, rest, and so on. The maximum angle of stability depends very much on the material: It is greater for angular grains than for spherical ones.

Our granular materials are dry. At our next destination, we will encounter pastes: wet granular materials with a dense concentration of particles. And here we are in Naples, our final station, which was also our starting point and where we were promised a miracle . . .

CHAPTER 14

LIQUEFACTION OF
THE "BLOOD" OF ST. JANUARIUS

I had read so much about the manifestations—both visual and spiritual—of the liquefaction of the "blood" of St. Januarius that in the end I knew only one thing: I had to head to Naples and see for myself on the Saturday before the first Sunday in May.

NAPLES, SATURDAY, MAY 3, 2008

I breathed in the landscape and now to draw it, hold my breath.

—Paul Claudel, *A Hundred Movements for a Fan*, 1927

Having arrived at the duomo at the crack of dawn, I observe the preparations of the town's minor patron saints for their annual outing: fifty-one impressive silver reliquary busts. Borne on the shoulders of parishioners from the neighboring communities and by young Neapolitans amid the cries, applause, and shower of flower petals tossed by people crowding into the streets and onto balconies, the saints will progress through the Via del Duomo, Via San Biagio dei Librai, Via Benedetto Croce, and finally enter the church of Santa Chiara. There, the city's archbishop will give a mass, during which the faithful will wait and hope for the liquefaction of the blood. "The procession has followed this route for only the last hundred years," Monsignore Vincenzo De Gregorio, a priest at the duomo and custodian of the relic, tells me later. "Before, some bearers used to duck out of the line for a few minutes to go take their saint on a tour of their favorite neighborhood!"

I return at 5 o'clock in the evening for the preparation of the relics in the Chapel of the Treasure, built in the seventeenth century inside the cathedral, in Neapolitan baroque style and dedicated to the cult of St. Januarius and the preservation of some of his relics. A priest dresses the saint's gold and silver bust with a miter, a cape, and a necklace encrusted with diamonds, emeralds, sapphires, and rubies, the work of three goldsmiths from Provence. In 1305 Charles II of Anjou, count of Provence and king of Naples from 1285 to 1309, donated it for the millennium celebration of the martyrdom of Januarius. The head contains a bag of relics. It so happens that it had been opened the day before my arrival for an inventory in the presence of a paleoanthropologist: "Earth, skull fragments, small hand and foot bones . . . and two insects," Monsignore De Gregorio tells me later with a quick smile.

The small chapel fills rapidly, and Italian and foreign TV crews close in to film the release of the reliquary. Instead of the "loud, ecstatic crowd" I was promised, I see only five *mammas,* real Neapolitan-style matrons, rushing forward as soon as the chapel doors open, determined to find seats in the front pew. With strong, confident voices, *i parenti de San Gennaro*[1] begin and end the preparatory ceremony with emotional songs and lamentations in the name of the saint. One hears clearly that they pray more fervently than the rest of the crowd. They will repeat the performance every day for the next week at the end of the evening mass at 6 o'clock, when the faithful are allowed to come and kiss the reliquary. Otherwise, I remark only a dense gathering of unobtrusive, meditative souls and tourists observing the ritual, some of whom smile mockingly. The reliquary is enshrined in a fortified safe behind the main altar. A priest disappears, then reappears with the reliquary, provoking a round of applause. The objects of interest are two vials, one large, one small, measuring 60 and 25 cubic centimeters, encased in a receptacle made of two thick glass plates. The entire assembly is framed in silver, with a handle. The large vial is two-thirds filled with "blood." The smaller one plays no role in the miracle; it contains only traces of material: "Practically dry, it looks like traces of earth," the monsignore will tell me later. The priest places the reliquary at the top of an ornate platform to be carried through the streets. The "blood" looks like a thick black paste, "hard as stone," according to a member of the deputation[2] eager to testify to its consistency as it is brought out of the safe. The ceremony in honor of St. Januarius has been going on since 1337. The liquefaction of the "blood" first occurred here in 1389. The church has never recognized it as a *miracle,* but rather refers to as a *wonder.*

The attentiveness and emotion of the Neapolitans in the streets are obvious. A security cordon attempts to control the convoy. Despite the dense crowed, I manage several times to get close to the reliquary, particularly when the procession, passing in front of a church, stops to allow a priest to come out and greet St. Januarius, swinging a censer. The reliquary gets bounced around a bit in the alleyways with their uneven cobblestones and by the hoards pushing to join the procession. Right up to the delivery at the church of Santa Chiara, I notice no softening of the material.

Toward 6:30 p.m. a long mass begins. During the service Sua Eminenza Reverendissima Cardinale Sepe, archbishop

On the way to the church of Santa Chiara in quest of a phase transition.

of Naples, reminds everyone that because the "blood" of St. Januarius represents a symbol of hope and life, his cult represents the exact opposite of the tradition of death and violence of the Camorra.[3] The first rows are occupied by city dignitaries and diverse congregants; from that point onward, I can see the reliquary only from afar. Cardinal Sepe holds it by the handle. He checks regularly to see whether the liquefaction in the bulb has begun by gently, slowly tipping the object, aided in his observations by Monsignore De Gregorio and the members of the deputation. "At 19:43," I read later in *Il Mattino*, "eighteen minutes after the beginning of prayers," Monsignore Sepe declares that St. Januarius's "blood" has liquefied. Count Augusto Cattaneo di Sannicandro smilingly waves a white scarf in accordance with an age-old rite that has the effect of breaking the surprisingly palpable tension among the attendants. Thunderous applause erupts in the church and continues for some time. Here, again, the reaction is not fanatical. The procession reassembles. During the return, I note that the contents of the vial are vaguely soft and gelatinous, very slightly liquid—which can be seen only by looking very closely. Moreover, one of the porters, during a pause in the procession, lowers and raises his shoulder to convince himself of the mobility of the sensitive material (he is right; nothing beats experiment). Finally, the reliquary is restored to the chapel, and the five *mammas* resume their singing and lamentations.

Monsignore De Gregorio, informed of my visit, had assured me that he would happily meet with me to answer my questions. I had clearly told him what I wished: to discuss the response of the faithful and his own attitude toward the event, what he thinks it means—and to talk about the *miracle* and other aspects of the cult. I had specified that I had no desire to be provocative vis-à-vis the explanation of the phenomenon. I wished only to obtain a firsthand account, and in any event, the context of my observations and the data available to me were not equivalent to scientific protocol. I had decided to present myself to the monsignore the following day, not wishing to impose on his time on a day that was already packed with activity. However, at the conclusion of the ceremony, lo and behold, he approaches the microphone: *Vorrei sapere se il Professore Mitov è qui?*[4] After an instant of surprise, I move toward him to indicate my presence. He gestures for the police to let me through, and, venturing into the space before the main altar, I realize that the priest is inviting me right there and then to examine the object close up and to assist in the ritual of returning it to the safe. Standing in front of me, the monsignore takes a moment to manipulate the reliquary, tipping it slowly back and forth. Remarkably, the dark, gelatinous fluid does not stick to the walls, for which it clearly has no affinity. "Sometimes the liquid sticks more; it is much more viscous," says the monsignore. At other times it is "like . . . mayonnaise," he adds sheepishly, for want of a better analogy.

The week following the *miracle,* the reliquary remains on display in the chapel, and every day, during the 6 o'clock evening mass, the faithful can receive a blessing through the touch of the object on their forehead. The priest does not permit them to lay their hands on it; if too fervent a believer reaches out a hand—which I saw happen once—the priest moves the reliquary out of the way. After that Saturday I attended the blessing three evenings in a row. Each time I saw the fluid unmistakably liquefied. The crowd was smaller than on the day of the ceremony and varied from day to day; sometimes only a handful showed up.

During the discussions that we had after the masses, I told the monsignore that I had been surprised not to see any show of fanaticism. I asked him whether this attitude changed when the liquefaction did not occur. "No, no fanaticism. You know, Neapolitans are very practical. If the liquefaction happens, they are reassured, happy—everything is going as they expected. If not, they are a little sad, of course, but it stops there and does not last." Monsignore confided that he himself was a little tense during the mass—up until the prayer that brought on the liquefaction. Not that he was worried about the consequences of nothing happening in the reliquary, he told me serenely, but because it would fall to Cardinal Sepe—who had been at his job for only two years—to find the appropriate words to soothe the Neapolitans. And, naturally, the monsignore hoped that his illustrious host would not have to do that on such a beautiful day.

I had read sensational and disturbing accounts claiming that the material, once liquefied, could boil, foam, turn brick red, and even change volume. But the monsignore had never seen such a thing. His predecessors, the abbots in charge of the chapel, kept a register dating from the seventeenth century in which they recorded the ceremony proceedings, the state of the fluid, and the visits of religious officials, sovereigns, and other celebrities to see the reliquary. So the monsignore told me in the sacristy, moving his hand in sweeping circles to indicate the library shelves lining the walls that contained these testimonies, which historians have apparently not yet mined. The monsignore confirmed that the state of the vial could be very different from one ceremony to the next. In September of the previous year, "the vial was already a little liquid when we opened the safe." In December the liquefaction failed to occur. The monsignore then confided to me something I had neither asked about nor expected: "I read an article on the work of a chemist, Garlaschelli, who fabricated a fluid that liquefied on agitation by shaking. I had a sudden doubt, and I shook the reliquary vigorously. But it had no effect." During our conversations and even without me evoking the hypothesis of a false relic, the monsignore calmly remarked: "Does the vial contain something other

On the evening of the procession, Monsignore De Gregorio carefully observes the liquefied state of the contents of the reliquary one last time before returning it to the safe until the following day. Years of manipulating the reliquary have done nothing to diminish his interest in and his wonder at the physical phenomenon.

than 'blood,' a substance capable of causing the change in state? I do not know."

SENSITIVE "BLOOD": IMITATIONS

Eusèbe Salverte (1771–1839) was a French politician who—between historical and political treatises—enjoyed writing about the occult sciences as a way of demystifying them. *Des science occultes ou Essai sur la magie, les prodiges et les miracles*,[5] published in 1829, contains a commentary on the Neapolitan miracle:

> In the present day, at an annual public ceremony at Naples, some [drops] of the blood of St. Januarius, . . . collected and dried centuries ago, becomes spontaneously liquefied, and rises in a boiling state to the top of the vial that [e]ncloses it. These phenomena may be produced by reddening sulphuric ether with orcanette . . . , and mixing the tincture with spermaceti[: t]his preparation at ten degrees[6] above the freezing point . . . , remains condensed, but melts and boils at twenty. To raise it to this temperature, it is only necessary to hold the vial which contains it in the hand for some time.

Pierre Larousse essentially copied and pasted this recipe into his *Grand Dictionnaire universel du XIXe siècle*. For him, the explanation is straightforward, and his judgment merciless: "Nothing is easier to do than this supposed miracle, which so vividly excites the enthusiasm of an ignorant population." Well! It would be hard to find a more contemptuous opinion. And to think that I hung around the dark corners of the cathedral for that instead of going swimming in Capri! Nonetheless, it is in fact true that the *Grand Dictionnaire*, written in an unfettered, at times disconcerting style, disseminated biting commentaries, prejudices, and scientific absurdities.[7]

So what is this recipe? It involves coloring spermaceti red using orchanet—a plant whose roots contain a dye—by diluting them in an appropriate solvent (such as sulfuric ether), which is then allowed to evaporate. Despite its etymology—early fishers believed the substance to be the animal's semen—it is a waxy white matter found in the head of sperm whales. Spermaceti is a lipid. Used in cosmetology, it was replaced by jojoba oil, whose chemical structure and properties are similar, which spared dolphins and whales a brutal and intensive hunt. The experiment was reproduced at the University of Nice–Sophia Antipolis, in France, by physicist Henri Broch and students. They tested several variations, including coconut oil, which is waxy below 24°C and oily above.[8] Tinted coconut oil is encapsulated in a glass tube, and liquefaction is produced by applying a hand to the tube provided the temperature of the room is below 24°C. Otherwise, the solid-fluid transition will already have taken place. This demonstration is useful because it reminds us that there are natural substances whose consistency is sensitive to temperature and in a reversible way.

However, in and of itself, the analogy would probably not be convincing to a visitor who had witnessed the miracle on May 3, 2008, and the daily ceremonies the following week because some of the elements of the scenario were missing. The presence of a crowd raises the temperature in the church (*Il Mattino* counted two thousand people at the mass), and the three ceremonies are in fact attended by an exceptionally large number of visitors. However, in the days that followed, while it was on display, the "blood" appeared constantly liquid even though the number of people in the chapel changed; in other words, the increase in temperature linked to the number of bodies varied from one day to another. The fact that, during some ceremonies, the "blood" liquefied as soon as the safe was opened and that the phenomenon was decidedly rarer in December (when the ceremony takes place in the duomo) argues for the thermal phenomenon. A simple test would be to record the temperature of the vial while observing the state of the fluid from the beginning to the end of the ceremony. As Broch says, "The correlation between heat and the lique-faction of the 'blood' should, at least, be considered," adding that one should not assume a link between cause and effect. To explain the observations of the liquid state from the opening of the safe would also require examining the walls of the compartment—is there an indirect source of heat, the mechanical action of a motor or ventilation? Luigi Garlaschelli,[9] a chemist at the University of Pavia and a member of the Italian committee for the scientific investigation of allegations of so-called paranormal phe-nomena, has proposed that, since Naples is a volcanic region, weak ground vibrations could trigger the lique-faction of a fluid at rest, provided it has *thixotropic* properties—from the Greek *thixis*, to touch, and *tropos*, changing—that is, transforming by touching. In addition, the liquefaction is even greater when the fluid is shaken during the ceremony.

Which brings us to the point. The idea of a substance with a consistency sensitive to shaking was initially pro-posed in 1890 by an Italian academic, Albini, and revisited in 1949 in a book by Alexander and Johnson on the science of colloids[10] that made specific mention of a "thixotropic gel"[11] with regard to the "blood" of St. Januarius. This idea was subsequently promoted in the 1990s by Garlaschelli, who made a thixotropic fluid with ingredients available in Naples in the fourteenth century (a time when alchemists mixed everything they could get their hands on, hoping somehow to produce materials with unlikely qualities): iron chloride (found in the stones around Vesuvius), calcium carbonate (eggshells, paint pigments from the Middle Ages), salt, and water. The one slightly technical step lay in dialyzing the mixtures, with the aim of electrostatically stabilizing the colloidal particles in water. At the Centre d'Élaboration de Matériaux et d'Études Structurales in Toulouse, France, we reproduced Garlaschelli's recipe. Within a few seconds of moderate agitation, the material indeed exhibited a nice transition from quite a thick mass to a very liquid fluid. Left to itself, the material resolidified in ten seconds or so. Qualitatively, the demonstration was dramatic and convincing. However, we are far removed from the time scale and details of the Neapolitan transition—in any event, the one I witnessed in May 2008.

COMPLEX FLUIDS: A FEAST OF SENSITIVITIES

A fluid whose viscosity increases with deformation is a *shear-thickening fluid*. If you mishandle pizza dough, it becomes rather tough, and you can make a ball out of it. If you lay it down, it flattens out. Silly Putty, which is made of silicone, has the same character: Molded into a ball, it bounces hard when thrown, but when laid on the ground, it stretches as far as it can and starts to shine, a sign of the absence of bumps. On the contrary, if the viscosity of the fluid diminishes as deformation increases, the fluid is a *shear-thinning fluid*, like paint or yogurt. A purely shear-thickening or -thinning fluid is one with no critical stress:

Everything Flows

The Greek philosopher Heraclitus of Ephesus was haunted by the transience of things. His words—*panta rhei*—express the constant nature of change in life and the universe: "You cannot step twice into the same river," he added. His thought also evokes the idea that every object contains within itself the seeds of its own negation. It is similarly tempting to conclude that the "blood" of St. Januarius in the solid phase contains—at the same time and in the same moment—something that will ultimately change it into a fluid. Heraclitus's words have been the motto of the American Society of Rheology since its founding in 1929. Rheology (from the Greek *rheo* [flow] and *logos* [study]) investigates the deformation of matter and its flow in response to mechanical stress. A force applied to a material implies a deformation proportional to the force exerted. The relation of the force to the deformation constitutes its viscosity. Accordingly, the viscosity of a fluid characterizes its resistance to flow.

Its character is expressed as soon as it encounters a stress, whatever its magnitude.

A *Bingham fluid,* in contrast, requires a minimum stress to flow. It is named after the American chemist Eugene C. Bingham (1878–1945), who made many discoveries in rheology. A Bingham fluid typically results from dispersion in a solvent of solid particles ranging in size from a few nanometers to several tens of micrometers. When the particles are sufficiently numerous, they touch, and the solvent is located in between: The liquid is pasty. Making it flow requires overcoming the forces between the particles in contact, which give the paste its cohesion. A minimal stress is necessary. Rheologists know that the idea of threshold is rather subjective because it frequently depends on the sensitivity of the apparatus used to measure it. An artist, painting with oil, has a specific requirement: to be able to re-work a layer applied on the canvas or, at some later time, to paint over it with a new color. If the latter, how does the artist keep the layers from mixing? By using "Flemish medium," a paste additive for paint that appeared in the seventeenth century. It enables retouching by refluidifying a layer of apparently dry paint. Paint containing Flemish medium is a Bingham fluid: It requires a minimal force with a brush to apply it. In addition, it sets up in a few minutes, taking on the consistency of a gel but without drying. It can be worked again at will, although over time it requires more force to apply. To cover a layer, the painter makes little dabs with a brush so as not to exceed the flow threshold of the paint the painter wishes to cover but sufficient to exceed the threshold required for the covering paint to soften. By the same token, if the painter wishes to mix both layers, it will be necessary to apply greater force to exceed both thresholds. Rubens (1577–1640) could well have painted *A Village Wedding* in a single day thanks to the Binghamian properties of his paint. Apparently, the exact composition of the Flemish medium of the time remains a mystery, but we do know that it contained linseed oil and resin. Dissolved in heated oil, the resin disperses in droplets, progressively forming a network, and the paint sets in several minutes. One must wait days, even months, for the paint to be entirely and definitively dry—the time needed for the painter to refine the creation.

A Bingham fluid is very often thixotropic: paper pulp, cement and mortar, mastic, food pastes, and cosmetics. A thixotropic fluid is a *yield-stress fluid*, such as a Bingham fluid, but this quality is not sufficient to characterize it: Its viscosity continues to decrease with time even as the stress remains constant. Thixotropic fluids include ketchup, the synovial fluid of joints (which require the right viscosity for the right movement), and ballpoint pen ink, which flows when a pen is pressed to paper but gels when the stress is removed. To a painter of buildings, the thixotropic properties of paint are a blessing: The paint adheres to the brush but spreads when the brush is dragged and remains on the wall at the conclusion of the motion. It smoothes under the effect of surface tension, but because of the interactions between the particles suspended in the fluid, its own weight is not enough to make it run. One way of imparting thixotropic properties to a paint that has none is to add fine sand to it. The behavior of quicksands, as in the bay of Mont Saint-Michel, is explained by the thixotropic character of clay in water. The recommendation that it is better not to thrash around in this sensitive matter is reasonable. However, I take issue with Hollywood screenwriters,[12] Sir Arthur Conan Doyle,[13] and Honoré de Balzac, who, as part of an engaging allegory, wrote that "the quicksands of the Loire never give up their prey"[14]: An animal or a human would never completely disappear in quicksand[15] because of the difference in density between the body and the sand . . . the saving thrust of Archimedes! In earthquake zones, clay soil will liquefy in response to the slightest seismic activity. A frequently cited example of a thixotropic catastrophe is that of Aberfan, in Wales, in 1966: Within a few minutes, the slide of a water-soaked heap of mine waste swept away houses and a school, claiming 144 victims, including 116 children.

The thixotropic behavior of a clay gel with just 4 percent of bentonite in water is easily shown by a wonderfully simple teaching experiment. Deposit a fistful of this

gel on an inclined plane covered with sandpaper to keep
the viscous mass on the surface. At a critical value of the
angle of inclination, the material will start to fluidify
when the flow stress equals the force of gravity per unit
of surface; it takes only a small decrease in the angle for
the gel to stop flowing. The viscosity of the gel decreases
gradually, which accelerates flow. That decreases viscosity even further, and so on. This avalanche phenomenon

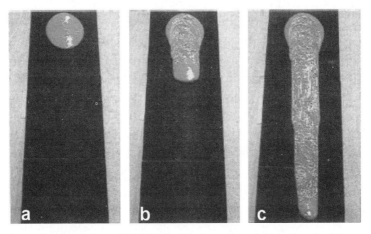

A very simple experiment highlights the thixotropy of a colloidal gel made of 4
percent clay in water (P. Coussot, Q. D. Nguyen, H. T. Huynh, and D. Bonn,
"Avalanche Behavior in Yield Stress Fluids," *Physical Review Letters,* vol. 88, no.
17, 2002, p. 175501–2). It reenacts on a small scale the dramatic liquefaction of
clayey soils in Canada and Scandinavia. (a) The gel is deposited on a plane tilted
at a precise angle. The experiment has just begun. The ball of matter retains its
shape. (b) An image taken after a few seconds: the gel has started to flow.
Gravity pulls the material downward. This movement creates mechanical
agitation, and the thixotropy is expressed. Reducing the angle by merely two or
three degrees would suffice to halt the phenomenon. (c) Several tens of seconds
afterward. Fluidization continues inexorably, the viscosity of the gel diminishes
with time, and its flow accelerates, culminating in an avalanche. The inclined
plane is two meters long, a distance it takes the fluid less than a minute to travel.

is remarkably similar to what occurs with granular matter (fresh snow, sandpiles) and gives rise to rapid fluid transport over great distances.

Let us look a bit more closely at the origin of thixotropy, which colloid scientists still find mysterious but which forms part of our everyday experience without our knowing it.

TRANSFORMATION BY TOUCH

Although thixotropic materials may appear to behave similarly, the small-scale mechanisms at work are different. For granular material dispersed in a fluid, networking of the particles explains its resistance to movement. Emulsions and foams are fixed by a network of liquid droplets or air bubbles. In a gel such as toothpaste, polymer chains and particles stick together. The relation between viscosity and structure is complex: Two samples of a single thixotropic fluid may have the same viscosity at a specific shear rate, yet be organized differently. This *structural hysteresis,* a product of the history of the sample, means that the return of a thixotropic fluid to its initial structure following fluidification does not occur via an inverse pathway. The case of paste is particularly interesting because its behavior resembles that of the Neapolitan fluid. A paste is a material made of one or several constituents dispersed in a liquid at high concentration. A continuous network of interactions between these objects propagates throughout the material. Given the variety of the constituents, the possibilities for internal structuring are enormous: diverse forms of solid particles (globules, needles, rods, platelets), frequently asymmetrical, of differing character (crystal, polymer, glass, surfactant, oil, water), all sizes (nanometer, micrometer, millimeter), and dispersed in an aqueous or hydrophobic matrix.

This range of possibilities explains the current lack of a unified scientific approach to describing pastes. Interactions between noncolloidal particles (larger than one hundred micrometers) differ from those between colloidal particles, bubbles, and droplets. Interactions occur between the dispersed constituents and the carrier medium and between the constituents themselves. The network of interactions evolves over time as the simple consequence of very small rearrangements of ingredients. Thermal agitation allows them to explore a variety of positions, but its role is limited because diffusion of the constituents is slowed down. The transition of fluid to paste is a *jamming transition*. The viscosity of a paste increases with resting time ("aging") and decreases with the time it takes to get it flowing ("rejuvenation"). The structure of a paste results from competition between the tendency of the system to return to its equilibrium configuration (restructuration) and its continuous breakage as a result of flow (destructuration).

The description of the transition between the solid and the liquid phases of paste is still a subject of debate. The transition is not abrupt since the applied deformation propagates heterogeneously throughout the medium— the tangled structure is not the same everywhere. A paste that is shaken may appear as clumps of thick material moving in a much more fluid medium. There are at least two explanations for this behavior: The distribution of shear is not the same throughout—there are *shear bands*— and the behavior of the paste near the walls of the receptacle reflects the very particular sliding of a thixotropic fluid on a surface . . . It is impossible to ignore these two experimental facts in thinking about the "blood" of St. Januarius on the evening of May 3, 2008, that is, the look of the gel, which appeared to be flickering, and its nonadherence to the walls of the vial. The minimum stress required for the solid-liquid transition is a function of not only the resting time but also the investigation time! The concept of viscosity is relative and depends fundamentally on the range of time over which the researcher studies the evolution of the material. The earth's mantle behaves like a solid over a human lifetime and like a viscous fluid from the perspective of geological time.

Thixotropy is a complicated phenomenon to model. For a few cases (foams, "simple" colloidal gels), methods exist for applying the rheological parameters of paste in the solid phase. Unfortunately, we cannot say the same for the liquid phase, and thixotropy remains the least well understood problem in the land of complex fluids.

THE "BLOOD" OF ST. JANUARIUS: A YIELD-STRESS FLUID?

If we were entrusted with the curious mission of fabricating a St. Januarius fluid, we might draw up the following list of specifications:

- Our material will be a colloidal fluid with several ingredients—different-sized particles dispersed in a liquid or even an emulsion (fat in aqueous fluid or the inverse; fat in resin or the inverse)—to ensure the versatility of the manifestations of the phenomenon over time and its sensitivity to context. The transition of the solid (or waxy) state to the gelatinous (or liquid) state will result from the decrease in viscosity of a paste containing several constituents whose mutual interactions will change and not from a phase transition in a single substance like spermaceti or coconut oil.

- A temperature threshold will be necessary to trigger the complete transition; a minimal viscosity will first have to be achieved by temperature rather than by some kind of mechanical jarring. This condition is connected with the behavior of the Neapolitan fluid during the three times of the year when the faithful pray for its fluidification: rather viscous in May, totally liquid in September (sometimes even as soon as the safe is opened), and laborious liquefaction, even failure, in December. But this condition will not suffice. The fluid will have thixotropic properties, for thixotropic fluids present flow scenarios that depend on previous processing, the stress applied (or the speed of deformation), and the time (duration of shear, resting time), which makes for a veritable feast of sensitivities. In short, thixotropic fluid is so much better! Once the minimal temperature has been achieved, any stress—even a weak one—can trigger fluidification. The chief function of this stress is not to increase the speed at which the temperature-induced change of state occurs but something more intrinsic; rather, its function is to enable the expression of thixotropy. The final state depends not only on temperature but also on the degree of agitation. There is certainly no shortage of terms for describing the fluid state: *liquido, tendere, congelato, duro, durissimo,* and *globo* or *globetto* if a hard core coexists within a soft medium.[16]

- The time to revert to the solid phase may be long, the time required for all of the constituents to regain a fixed structure in a very viscous environment. If the fluid experiences even a mild shaking during this period (manipulation during a daily mass, for example), the restructuration time will be prolonged, and the moment of solidification delayed.

The crucial role of shearing added to thermosensitivity may explain why the scenarios reported are so different: resolidification possible during the procession (once the reliquary is taken out of the cathedral, the temperature becomes too low for the fluidification, which was activated by shaking, to persist); persistence of the gel state although the same temperature on another day might be sufficient for liquefaction (because the initial internal structure varies from one ceremony to another, the initial state for the following ceremony is different as well and, consequently, the specific temperature threshold); the versatility of the stress threshold for fluidification, and so on.

NATURAL VERSUS SUPERNATURAL

The astrophysicist Carl Sagan used to say, "Extraordinary claims require extraordinary evidence." Is the behavior of the Neapolitan fluid out of the ordinary? Certainly not, since thermosensitive and thixotropic colloidal fluids are known. Fluids resulting from recipes published in the literature must be understood as model systems; their authors do not pretend to have produced *duplicatas*. Creating a faithful replica (no pun intended) would require establishing a record of all of the manifestations of liquefaction—that is, if we could manage to assemble all of the documents containing dispassionate, reliable, and precise testimony, which is no small thing. The hypothesis of a chemical recipe is a plausible one. It is not necessary to invoke a miracle. The "supernatural" hypothesis may thus be considered as superfluous. Does it mean that it is false? Certainly not. First of all, only a sample sent for analysis to different laboratories, according to a strict protocol, would establish the composition of the "blood"; indirect methods[17] that did not involve opening the vial are equally possible.

Next, assuming sampling would be permitted, the objectives of the analysis would have to be made clear. Even though the fluid would be a product of the laboratory, why make a point of proving that what happens during the ceremony is not essentially spiritual? This enterprise makes no sense to a scientist. The church at Argenteuil, France, possesses a tunic supposedly worn by Christ during his ascent to Calvary. Carbon-14 dating has shown that it does not come from the first century. As expected, some people refuse to acknowledge the findings. However, Father Rosier, a former priest of the church, says, "What is the point of criticizing scientific analysis? All this is ideology, and I trust the dating. That the relics are real is secondary to me. What matters is that these objects help us to live in faith."[18] Providing an alternative to the "supernatural" hypothesis is what motivates scientists such as Broch and Garlaschelli. Broch heads the Zetetic Laboratory,[19] a center of research and information on so-called paranormal or extraordinary phenomena,[20] unique in France, and summarizes his approach as follows: "The right to dream has a counterpart: the duty of vigilance." In 1993, at the School of Science at Nice, he instituted the first courses in zetetics, the goal of which is to understand the scientific method.

The phenomenon of liquefaction of the "blood" is not used (or is no longer used) to manipulate the crowd and does not constitute (or no longer constitutes) the sine qua non of devotion to St. Januarius. The success of the phase transition is "only" the high point of a ceremony to which many Neapolitans are very attached as a means of communing with each other and expressing their identity. Aside from the renewal of a secular tradition, the cult of St. Januarius represents stability and unity for the city.

The composition of the fluid in Naples is a mystery, and so is its origin. Does the fluid date from the era of sampling by the old woman we encountered at the beginning of this story? Or is it contemporaneous with the first occurrence (fourteenth century)? The vial could be a fake; it would not be the first time. For fake relics enjoyed a brisk traffic during the Middle Ages, when the cult of the saints reached its peak. In his *Dictionnaire critique des reliques et des images miraculeuses* (1821), writer and freethinker Collin de Plancy counted eight arms for Saint Blaise, thirty-two fingers for Saint Peter, eleven legs for Saint Matthew, ten heads for Saint Léger, and three bodies for Saint Agnes.[21] Was the fluid replenished? And is it still from time to time? Broch, in the interest of thoroughness (and he is right), reports that a witness claims to have seen "a young acolyte from the cathedral of San Gennaio"[22] order a miraculous fluid from a pharmacist in Naples at the beginning of the last century (whose blatant lack of discretion would be surprising, by the way). As with the account of the blood boiling, I cannot resist quoting an adage from old Roman law: *Testis unus, testis nullis* (One witness is no witness).

A Plea to Restore the Forgotten Miracles of Provence

In fact, Italy is full of relics of the sensitive, centuries-old blood of Christ and of the saints—San Pantaleone, San Lorenzo, San Giovanni Battista, San Stefano, Santa Patrizia, Santa Chiara da Montefalco, and so forth—and they are scattered throughout the rest of Europe (Auvergne, Spain, Belgium, Ireland) as well. However, these *miracles* were kept cruelly in the shadows by the popularity of the cult of St. Januarius. A sampling of the Provençal marvels of yore includes the following: Salverte reports that "[i]n Provence, in the 16th century, when a consecrated vial, filled with the blood of St. Magdalene, in a solid state, was placed near her pretended head, the blood became liquid and suddenly boiled."[1] It is possible that this scene unfolded at Saint-Maximin-la-Sainte-Baume, in the Var, where Mary Magdalene died, according to Provençal tradition. A repentant sinner who found Christ and sole witness to his resurrection, she spent her long years of repentance in the grotto of Sainte-Baume (*baoumo,* grotto in the Provençal dialect) after having landed in Saintes-Maries-de-la-Mer on the way from Palestine and evangelized Provence. In 710, her remains were hidden in a crypt to protect them from the threatened depredations of the Saracens. She was found nearly complete on December 9, 1279, during a search organized by Charles II of Anjou himself "with a determination and ardor equal to his piety, to the point where he was soaked with sweat."[2] One can easily imagine him resolutely intent on not giving any workmen the credit for having discovered the relics. The current basilica was erected in 1295 on this spot according to plans drawn up by Pierre d'Angicourt, who also created the Castel Nuovo at Naples. Yes, Charles II of Anjou de Saint Maximin is the very same king of Naples who in 1305 donated the bejeweled bust of St. Januarius. It is a disturbing coincidence, perhaps fortuitous, but one I would like to point out. It is worth digging a little to discover whether a link exists between the liquefactions of Var and Naples and, if so, to return this forgotten miracle of Provence, unjustly deprived of the longevity of its Italian counterpart, to its rightful place. The fascinating *Guide de la Provence mystérieuse*[3] tells us that also at Saint Maximin during Charles's searches, a vial was found that contained fragments of a puzzling red material—stones stained with the blood of Christ that Mary Magdalene collected at the foot of the cross.[4] Every Good Friday, after the reading of the Passion, these ordinarily dark red stones would turn a vivid vermilion, and the blood would liquefy and boil, rising and falling in the vial.[5] A long time ago, this miracle attracted up to six thousand pilgrims![6] Alas, the said holy vial,[7] which Mary Magdalene always kept with her, was stolen in April 1904 when the basilica was looted.[8] Let us fervently hope that this ultrasensitive material remains, intact and functional, locked away in the care of some impious collector or antique dealer and continues to astound its private observers—perhaps reunited (who knows?) in a secret society. If its possessor reads these pages, that person should know that it is never too late to choose the path of redemption and to discreetly approach the priest of the basilica, whose capacity for forgiveness knows no bounds. In any event, what a shame that the inhabitants of Saint-Maximin have been unable to continue to celebrate and preserve these miraculous events—though happily for the Neapolitans—compared with which the sometimes timid liquefaction of the blood of St. Januarius might seem a pale second! However, to be fair, even at the cost of our own pride, the final score of the match between France and Italy is an incontestable 0 to 1.

Notes

1. E. Salverte, *The Occult Sciences: Philosophy of Magic, Prodigies, and Apparent Miracles,* trans. A. T. Thomson, New York: Harper and Brothers, 1847, vol. 1, p. 286.
2. Abbé Étienne-Michel Faillon, *Monuments inédits sur l'apostolat de sainte Marie Madeleine en Provence et sur les autres apôtres de cette contrée, saint Lazare, saint Maximin, sainte Marthe et les saintes Marie Jacobé et Salomé, etc., etc.,* J.-P. Migne, éditeur, Paris, 1848, vol. 1, p. 873.
3. Collectif d'auteurs, Tchou éditeur, 1965, p. 447 (reprinted regularly).
4. According to l'abbé Faillon (*Monuments inédits,* p. 914), these relics "were the object of a public, constant, solemn following, known to the papal sovereigns who resided at Avignon and even authorized by the practice of all the popes who made pilgrimages to Saint-Maximin."

5. Is this event distinct from the first, or was Salverte confused?
6. Reverend father Vincens Reboul, *Histoire de la vie et de la mort de Sainte Marie-Magdeleine,* imprimerie de C. Brébion, Marseille, 1682, p. 13.
7. Not to be confused with the sacred vial at Reims, which contained the holy oil used to anoint the kings of France during their coronation.
8. J. M. Béguin, *La Madeleine,* H. Aubertin et G. Rolle libraires-éditeurs, Marseille, 1905, p. 563.

Even if the event is true, how can we prove that the witness was not the perfect victim of a playful setup, a Neapolitan tall tale? Nor did he report witnessing the ultimate use of the substance.

PREMEDITATION VERSUS CHANCE

With the chemical hypothesis, the intent to trick is, curiously, almost always suggested as an explanation. Yet chance fabrication cannot be ruled out.

One might wonder about the exceptional conservation, over a similarly exceptional span of time, of blood purported to be pure, especially given that the vial has not always been depicted as sealed over the centuries. Without necessarily maliciously intending to create a relic having remarkable physicochemical behavior, the handler could have mixed the blood with a substance capable of preserving it[23] (balsam extract, gum, or aromatic resin), which would result in a thermosensitive and thixotropic emulsion or gel whose surprising behavior would have inspired him, subsequently, to attribute the result to a miracle. These substances had been used for embalming since ancient Egyptian times, which would explain the "divine odors," "celestial odors," or other "saintly odors" described by witnesses with regard to the opening of sarcophagi or vials containing the relics of certain saints and other blessed personages. Such was also reported of Mary Magdalene.

What if we decide in favor of the legend, which does contain fragments of reality, and what if—just to speculate—instead of sponging it away, the old woman had gathered up the dark dirt of Pouzzoles[24] (the site of the martyrdom), wet with the saint's blood, and, thus, being a great experimenter, created a colloidal fluid with complex behavior? So complex, in fact, that to this day, this research area—yield-stress fluids—remains a frontier in laboratories that specialize in the chemistry and physics of colloids.

That is it in a nutshell: the liquefaction of the "blood" of St. Januarius, or when history, legend, belief, faith, politics, and science assemble around the same table.

EPILOGUE

At the beginning of the evening I left Monsignore De Gregorio in the square of the cathedral where he had accompanied me, attentive and affable to the end. We exchanged a few more words about the Isle of Capri, where he had been born. I strolled pensively through the streets of the Spaccanapoli, my head filled with the chants of the descendants of St. Januarius:

> *Potenzia di San Gennaro, pruteggetece,*
> *Sangue di San Gennaro, defendetece.*
> *Miserere! Miserere! So' 'e peccate, so' 'e peccate!*
> *San Gennaro miserere.*

> Power of St. Januarius, protect us,
> Blood of St. Januarius, defend us.

Have mercy on us! Have mercy on us! For we have sinned!
For we have sinned!
St. Januarius, have mercy on us!

As night fell, fireworks crackled merrily in the distance. But it has been a long time since anyone danced a tarantella in the streets of Naples on the evenings of the great liquefaction, which really is a shame.

BONUS TRACKS

Let me leave this sensitive subject with a few brief reflections on the classification of the states of sensitive matter within that of matter generally.

BONUS 1: TOUGH MATTER TO CLASSIFY?
How We Perceive Matter

Our everyday experience brings us closer to liquids and solids than to gas, for we can touch and define them. If melting ice is a clearly visible manifestation of the transformation of matter, we equate the evaporation of water to its disappearance, pure and simple. It is the knowledge of the organization of matter on the small scale that leads us to group liquids and gases together as a class—fluids—markedly distinct from solids. However, for some time now, we have had to move beyond the duality of liquids and solids: The study of glass, to cite just one example, has shown us that, although it is solid, glass has no crystalline structure—notwithstanding its being called Bohême or Baccarat crystal. It is composed of liquids of extreme viscosity that would flow if the observer were willing to take the time—eons!—to investigate them. Crystal, glass, wood, and plastic are all called solid materials, but they have very different malleability and hardness and differ in their processing.

Classification Revisited

We are taught that matter exists in three fundamental states: *crystalline solid, liquid, and gas.* A crystal is a perfectly ordered arrangement of atoms or molecules. In table salt, chlorine and sodium atoms alternate with perfect symmetry. They have an extremely limited range of movement—at the most a few vibrational movements around very precise positions due to thermal agitation. The atoms and molecules of a fluid, on the other hand, have no particular arrangement or orientation and are very mobile (a molecule in a glass of water changes its place a hundred billion times per second). This is the reason a fluid fits the shape of its container. What distinguishes liquids and gases is the degree of this mobility. Liquids are very dense (difficult to compress), which forces their atoms into constant collisions and interactions with each other, which explains the changes in position and random movements. In a gas, on the other hand, an atom can travel a great distance before bumping into another atom: They are totally independent. That is why a gas is volatile and in some sense may be seen as an extended liquid. A gas tends to fill the available space and can fairly easily be compressed. The distinction between liquid and gas is thus by nature quantitative, whereas that between crystalline solid and fluid is qualitative, with reference to the order-disorder dichotomy of the arrangement of atoms and molecules.

Crystalline solid—liquid—gas: Where in this tidy system do we classify the members of the family of sensitive materials? And must this question, however legitimate, constitute a precondition? Although not often portrayed in the media, researchers' daily work consists in painstakingly defining the conditions for validating their experiments and calculations, not in proclaiming absolute, universal truths about the laws and behavior of nature. In fact, it is much more interesting and relevant to try to clarify the ways in which the characteristics and properties of a liquid crystal, emulsion, foam, or plastic differ from those of a crystalline solid, liquid, or gas than to seek to place any of these materials in one category or another.

Sometimes authors classify liquid crystals as a *fourth state of matter.* This runs counter to one of the paramount teachings of the science of liquid crystals, which is to discourage any attempt to classify or number the states of matter. *Solid, liquid,* and *gaseous* states . . . this classification is no more than a residue of bygone schooldays. It is more interesting to analyze how a liquid crystalline state manages to elude any kind of rigid approach than to try to categorize it one way or another and certainly not to create a fourth category.

Our exploration of the forms of sensitive matter was not conducted within a rigid framework of order and disorder or of continuity and discontinuity. That approach implies a crude division, devoid of nuance, that might have led us to believe that the transition of one state of matter to another—caused, say, by a change in temperature—is necessarily continuous or discontinuous but in no case gradual. The qualitative distinction between crystalline solid and liquid is irrelevant for sensitive matter.

Building Blocks, of Course, but You Have to Connect Them

Briefly put, the classical definition of the states of matter is a description of their atoms: The starting conditions are defined by their mutual arrangement. Reducing matter to its elementary constituents is a customary approach in science. However, stopping there is problematic as soon as the question of practical applications arises. Why does the volume of water increase when it becomes ice at 0°C? Most other liquids contract when they solidify. A model limited to describing a water molecule as composed of two hydrogen atoms and one oxygen atom rules out any answer to this question, which is far from trivial. It is also not benign, for if a facetious—"diabolical" might be a better word—operator were to discover an interrupter capable of reversing the tendency, the earth's ice floes would fall to the bottom of the oceans, which themselves would turn into a solid mass.

The analogy between a brick wall and matter consisting of atoms underpins the atomist view. Bricks are elementary objects, and their arrangement takes into account the global and macroscopic properties of the material. But a wall also contains cement. Accordingly, describing the building means being aware of the cement: Is the cohesion between the bricks weak or strong? If you lean on the wall, will it deform? Also consider its behavior in response to temperature—does the wall warp during the transition from winter to summer? And so forth. Although practical experience leads us to think of matter in terms of its individual components, we are not so quick to perceive the bonds that ensure their cohesion. Yet, no one disputes the fact that a living creature cannot be reduced to its constitutive organs, which do not function autonomously. Behind this remark is one of the reasons that the complexity of living matter is totally unlike that of inert structures. Although a hydrogen atom may be the same in any physics or chemistry laboratory the world over, that is hardly true of a living cell being studied in Boston, Gothenburg, or Tsukuba. Even a simple water molecule requires careful thinking about the structure's effect on the individual atoms. The identity of a lone hydrogen atom does not change as a function of the orientation in space from which it is examined. In the parlance of physicists, it is *invariant with respect to rotation.* However, as soon as two hydrogen atoms combine with one oxygen atom to produce a water molecule—in its familiar compasslike configuration[1]—the physicist and the chemist treat it differently: Each hydrogen atom becomes attached to a sign that indicates its special connection with oxygen. A hydrogen atom that is part of a water molecule can no longer be perceived as being the same as when it stood on its own. The compound modifies the properties of the components.

BONUS 2: DEFINING MATTER
Bonds: The Essence of Sensitive Matter
In humankind's quest to gather, classify, define, and name the objects of its discoveries, the states of sensitive matter are often treated as having a common denominator: the bond. It is here that the energy of interaction between molecules is less than that at play in thermal agitation (or at least comparable). Energy is a very abstract concept for

quantifying very concrete interactions between objects on all scales—from our macroscopic world to that of the basic elements of matter. Energy can be broken down as the sum of two expressions: *kinetic energy*, associated with the movement of atoms or molecules (or larger objects), and *potential energy*, resulting from the interaction of matter with the external world—through gravitation or an electric or magnetic field—or interactions between molecules, between atoms, and so forth. If the energy of interaction between the molecules that make up sensitive matter is of the order of that of thermal agitation, a small variation in the equilibrium temperature of the material may suffice to modify a property.

This has nothing to do with the "butterfly effect,"[2] which indicates a system's sensitivity to initial conditions. Nonetheless, the adage "small causes have large effects" is perfectly appropriate for sensitive matter when applied to a pile of grains or a gel. In this context, sensitive matter is a superb opportunity to study the chemistry and physics of instability.

Words

> I never quarrel about a name, provided I am apprised of the sense in which it is understood.
>
> Pascal, *The Provincial Letters* (1657)

What we call "sensitive matter" is known widely under another name, "soft matter." Pierre-Gilles de Gennes recounted that this expression first appeared in his group at the University of Orsay, France, around 1970, where it had been tossed out in jest by physicist Madeleine Veyssié because of its double (scatological) meaning:[3] "It was a 'private joke' within a small group . . . (only a handful of us were working in the area)," states Madeleine,[4] remembering "a happy atmosphere where everything was easy and got done very quickly." She added with wry humor and the modesty well known to her colleagues and friends: "In hindsight, I consider it my greatest—and only—achievement of note!" Although the coarse origin was quickly forgotten, the English version of the term stuck. For "soft" should indeed be understood as "labile" and "sensitive" rather than as an assessment of its consistency.

Americans frequently prefer the expression "complex fluid" to "soft matter." However, teachers hesitate to use "complex fluid" for fear of deterring students—probably a general precaution considering the growing scarcity of candidates willing to undertake lengthy studies and a dissertation in the exact sciences. And after all, the notion of complexity is even more elusive than softness: Understandably, all of the subjects of modern research—since they are young—are by nature complex. Many of their behaviors are thus unknown, and describing them is an ongoing process. Moreover, certain forms of sensitive matter are not *fluid* at all.

As for the word "fragile," it is confusing. Bluntly put, it evokes the character of a material like glass, which, once broken, cannot be repaired or reassembled—how different from a cellular membrane!—and does not at all suggest the softness of rubber and its resistance to tearing. Polymers belong to the family of soft or fragile matter, but some of them are as strong as steel. The natural composite we call bone—another member of this large family—possesses a fragility that is decidedly relative, fortunately.

NOTES

PROLOGUE

1. A. Dumas, *Le Corricolo*, Desjonquères, 2006.

**1. PEACEMAKING AMONG ENEMIES . . . EASY
WHEN A MEDIATOR IS INVOLVED**

1. The approximate composition of a chicken egg yolk: 50 percent water, 32 percent lipids (triglycerides, lecithin, cholesterol), 16 percent proteins, 2 percent other substances (vitamins, minerals).

2. Variable-sized sacs, typically between 0.1 and 100 micrometers in diameter. A micrometer is one thousandth of a millimeter.

3. Electrically charged atoms or molecules.

4. Molecular gastronomy is a fairly young scientific discipline that examines the phenomena at work during cooking. It produces knowledge, not delicacies. *Molecular cooking* is an application of the field. Hervé This, who coined the term together with Hungarian physicist Nicholas Kurti, insists on distinguishing the two: "Pierre and Marie Curie

helped to explore the structure of atoms, but they are not responsible for Hiroshima. Similarly, molecular gastronomy is not responsible for the results, be they good or bad, of the cooks who are part of the molecular cooking movement." See This's website: http://www.inra.fr/la_science_et_vous/apprendre_experimenter/gastronomie_moleculaire.

2. DISSOLVING FAT IN WATER: A QUESTION OF ORGANIZATION

1. D. M. Small, M. Bourges, and D. G. Dervichian, "Ternary and Quaternary Aqueous Systems Containing Bile Salt, Lecithin, and Cholesterol," *Nature*, 1966, 211 (51), pp. 816–818.

2. Positively charged ions.

3. E-mail from Small, July 27, 2007.

3. DON'T MIX, ASSOCIATE!

1. Extract translated from German to English, appearing in H. Kelker, "History of Liquid Crystals," *Molecular Crystals and Liquid Crystals*, 1973, 21, pp. 1–48, extract appears on p. 7.

2. A fatty membrane that isolates each nerve of both the brain and the spinal column, like a plasticized sheath wrapped around an electric wire. Myelin speeds the propagation of messages between the central nervous system and the rest of the body.

3. Bulletproof vests, motorcycle helmets, sailboats, chassis (with glass or carbon-reinforcement fibers).

4. G. Friedel, "Les états mésomorphes de la matière," *Annales de Physique*, 1922, 28, pp. 273–474.

5. F. Grandjean, *Autobiographie scientifique* (1937), available at the *Annales des Mines* website: http://www.annales.org/archives/x/grand jean.html.

6. The shattering of a crystal consists in its fragmenting into perfectly clean, planar sheets (which is possible only with certain crystals, such as gypsum or graphite). The crack is produced where the bonds between atoms are the weakest. Mica, which is organized in sheets, cleaves easily using a fingernail.

7. F. Grandjean, "Sur les franges d'interférence développées par le frottement et l'électricité dans certains liquides anisotropes," *Comptes rendus de l'Académie des sciences*, 167, 1918, pp. 494–496.

8. C. Mauguin, "Orientation des cristaux liquides par les lames de mica," *Comptes rendus de l'Académie des sciences*, 156, 1913, pp. 1246–1247.

9. Facsimile of the letter in Y. Bouligand, "Les cristaux liquides à la 'Belle Epoque,'" *Bulletin de la Société française de Physique*, 40, 1981, pp. 9–13.

10. The nanosciences deal with the manipulation of matter atom by atom, molecule by molecule. In actual fact, these days this definition is extended to "small" objects, typically up to a hundred nanometers, which may comprise billions of atoms.

11. For example, the intermediary scale (between several tens of nanometers and several tens of micrometers) between the molecule and the cell (or biological tissue) is where major biological events, such as the evolution of a tumor, communication between neurons, and cellular adhesion, take place.

12. J. Friedel, *Graine de Mandarin*, Odile Jacob, 1994, pp. 109–112, and correspondence dated July 5, 2007, to the author.

13. I beg your indulgence in imagining what Miss Friedel might have said at this informal moment in the history of science.

14. In "Montesquieu," *Tableau de la littérature française*, Gallimard, 1939, vol. 2.

15. Temperature is the translation of the movement of molecules: The higher the temperature, the more agitated the molecules. The most active molecules transfer their kinetic energy to less active ones.

16. In *Essai de critique indirecte*, Grasset, 1932.

4. RUBBER: A STORY NEARLY CUT SHORT

1. This is what we mean by the word "serendipity."

2. Kurt Vonnegut, *Cat's Cradle*, New York, Holt, 1963.

3. B. Derjaguin, "Polywater Reviewed," *Nature*, 1983, 301, pp. 9–10.

5. THE FIREFIGHTER'S JET STREAM: REACH FOR THE SKY

1. One part per million (ppm) is equivalent to one millionth—a milligram per kilogram, for example, or one grain of sugar in a bag that contains one kilogram of the substance. Often a percentage is substituted for ppm units to indicate a weak concentration—one milligram in one liter of liquid, say, although the liter (a unit of volume) should not be confused with the kilogram (a unit of mass).

8. BREATHING: AN UNSEEN TRIUMPH

1. By analogy, think of what happens with a soap bubble that is insufficiently inflated: A protuberance begins to appear in the film formed

on the ring dipped in soapy solution, but if the blower doesn't persist, the film goes back to its flat, stretched form.

2. When two phases (solid, liquid, or gas) are in contact, the pressure exerted by one on the other represents the force applied per unit of surface. The surface in question is the interface between the two phases.

9. FAMILIARITY AND DISTANCE: COLLOIDS

1. British physicist Thomas Graham introduced the term "colloid"—from the Greek, meaning "like glue"—in 1861. He wished to distinguish substances that, in solution, are capable of crossing a parchment membrane from those that cannot. Certain substances (sugar, mineral salts) systematically showed crystalline features, and the corresponding solute passed through the membrane, whereas albumin and gelatin were gluey. Based on these diffusion criteria, he proposed to class chemical substances in two major categories: crystalloids and colloids. This confusing and inappropriate classification was dropped in the early 1900s.

2. Assume a spherical grain of radius r. Its surface is $4\pi r^2$ and its volume $4\pi r^3/3$. The surface-to-volume ratio is thus $3/r$. If r is equal to 10 nanometers, 1 cubic meter of grains contains 300 million square meters of surface, or the equivalent of 50,000 football fields!

3. In the formation of cream from milk, fat particles migrate to the surface of the aqueous solution and form large floccules—that's the *fleur du lait*. Gathered with a strainer and dropped gently into a mold, it turns into cheese—Corsican *brocciu*.

4. The typical composition of fresh cow's milk is 86 percent water, 5 percent lactose, 4 percent fat, 4 percent protein, and 1 percent salt.

10. SENSITIVE COOKING

1. Visit of February 20, 2008.

2. Many cooks abandoned gelatin during the "mad cow" scare, and the popularity of vegetable-based gels skyrocketed.

3. Bonds can also be established through small molecules or ions that attach to certain macromolecular groups. For example, alginates gel in the presence of a tiny quantity of calcium ions.

4. The absorption capacity is less for urine because it is loaded with ions.

5. S. Galam, "Terrorism et percolation," *Pour la Science*, 306, 2003, pp. 90–93; and S. Galam, "The September 11 Attack: A Percolation of

Individual Passive Support," *European Physical Journal B,* 26, 2002, pp. 269–272.

12. PUTTING DRUG DELIVERY ON CONTROLLED RELEASE

1. The famous "Nothing is lost, nothing is created, everything is transformed," attributed to the chemist and philosopher Antoine de Lavoisier (1743–1794).

2. R. Langer, "New Methods of Drug Delivery," *Science,* 249, 1990, pp. 1527–1533.

3. U. O. Häfeli, "Magnetically Modulated Therapeutic Systems," *International Journal of Pharmaceutics,* 277, 2004, pp. 19–24.

14. LIQUEFACTION OF THE "BLOOD" OF ST. JANUARIUS

1. So-called because, according to tradition, they are descended from the saint's wet nurse.

2. The lay deputation of St. Januarius consists of twelve representatives—ten from Neapolitan nobility and two commoners—assembled under the direction of the mayor of Naples, an ex officio member. The deputation's mission is to help administer the Chapel of the Treasure in maintaining the cult of St. Januarius and to serve as an intermediary between the church and the city. The members of the deputation are passive observers during the ceremony and the display of the reliquary. The safe has two locks: One key is kept by Monsignore De Gregorio, and the other by a deputy.

3. The criminal underworld of Naples.

4. "Is Professor Mitov here?" In Italy, a researcher or teacher-researcher is de facto called "professor" as a sign of respect.

5. E. Salverte, *The Occult Sciences: Philosophy of Magic, Prodigies, and Apparent Miracles,* trans. A. T. Thomson, New York: Harper and Brothers, 1847, vol. 1, p. 287.

6. Centigrade.

7. . . . when he was not, depending on the subject, being purely sexist or racist. Taking the contemporary context into account does not explain everything. In its time, this dictionary was an immense literary success and an undisputed cultural reference for generations. See *Que faire des crétins? Les perles du Grand Larousse,* P. Enckell and P. Larousse, Seuil, 2006.

8. This recipe is older and was reported before the nineteenth century. In *Philosophoumena* (third century), the author—whose identity is

uncertain (either Hippolyte of Rome or Origen of Alexandria, both of whom were theologians)—writes about tricks used by Babylonian priests to impress the crowds: "Wax mixed with orchanet, as well as wax mixed with incense, produces a bloodlike liquor" (Henri Broch, pers. comm., Dec. 1, 2008). In the satire describing his voyage from Rome to Brundisium, the poet Horace (68 BC to 8 AD) also recounts impishly:

> Weather the next day
> improved, but roads were worse along the way
> to Barium, a walled-in fishing town.
> Once we departed, we continued down
> the Gnatia, constructed to annoy
> the water-nymphs, where we would all enjoy
> some jokes and snickering at the expense
> of locals telling us that frankincense
> melts on their temple steps without a flame.
> The Jew Apella may believe this claim;
> not I, for I have learned that deities
> pass through their time without our miseries
> and Nature's miracles are not hurled down
> when gods above us in the heavens frown.

(Horace, *The Satires of Horace,* trans. A. M. Juster, Philadelphia: University of Pennsylvania Press, 2008.)

9. Http://sites.google.com/site/luigigarlaschelli/.

10. A. E. Alexander and P. Johnson, *Colloid Science,* Clarendon, Oxford, vol. 1, 1949.

11. The generic term "thixotropy" made its appearance in the scientific literature in the 1920s.

12. *Indiana Jones and the Kingdom of the Crystal Skull,* directed by Steven Spielberg (2008).

13. *The Hound of the Baskervilles* (1902).

14. *Lost Illusions* (1836–1843).

15. Not all quicksand is composed of thixotropic clay. There are wet quicksands (traversed by water currents) and dry ones, where air circulates among the grains like those that trapped Peter O'Toole's young companion in David Lean's 1962 film, *Lawrence of Arabia.* This association of sand and water or sand and air causes the fragile soil to sag under sufficient pressure.

16. H. Bender, "Le miracle du sang de saint Janvier," *Étonnante parapsychologie,* éditions CAL, 1977, pp. 125–149.

17. Indeed, indirect investigations using spectroscopy were performed in 1902 and 1989. The idea was to study how visible light interacts with matter and provides information about its composition and structure. Aside from the fact that the work was never published in scientific journals (that is, scrutinized by peer review), the analyses are not conclusive. Garlaschelli questions the reliability of the instruments and the experimental protocol and points out that the two series of experiments reported the presence of hemoglobin but only as a partial ingredient. The authors themselves acknowledge that a red dye would have given the same result.

18. H. Ratel, "Jésus, ou comment faire du neuf avec du vieux," *Sciences et Avenir*, no. 710, 2006, p. 19.

19. From the Greek *zêtêtikos*, "he who searches for truth." Zetetician was a name given to the Greek skeptic philosophers.

20. Http://www.unice.fr/zetetique/.

21. C. Portier-Kaltenbach, *Histoires d'os et autres illustres abattis*, Jean-Claude Lattès, 2007, p. 14.

22. H. Broch, "La miraculeuse liquéfaction du sang de St Janvier," *Au coeur de l'extraordinaire*, Éditions book-e-book, 2004, pp. 311–313.

23. P. Saintyves, *Les Reliques et les Images légendaires*, Robert Laffont, 1987, p. 911.

24. *Pouzzolane* is the product of volcanic activity. In the form of fine red or black sand, it is used in certain cements and heat-resistant materials.

BONUS TRACKS

1. The hinge of the compass is occupied by an atom of oxygen, and each of its points has a hydrogen atom. The angle of opening is 104.5°.

2. In 1972 meteorologist Edward Lorenz, who had a finely honed (if obscure) sense of metaphor, gave a talk to the American Association for the Advancement of Science, titled "Predictability: Does the Flap of a Butterfly's Wings in Brazil Set Off a Tornado in Texas?" The expression, repeated and exaggerated, has since grown legs. Its use in the popular press, movies, and songs is nearly always opposite to its intended meaning: It is not the flapping of a specific butterfly's wings that triggers a natural catastrophe; any other winged creature could do the same, and the flapping could just as well prevent a tornado. Even the science behind the notion is arguable, for the energy the butterfly exerted on such a small scale would be dissipated before it could produce an effect of such magnitude. More seriously, chaos theory—whose (relative) popularity derives

from the butterfly effect—states that a small variation in a parameter inherent to a system can have a determining influence on the state of the system at a later date.

3. "Soft Matter: More than Words," *Soft Matter,* 1, 2005, p. 16.

4. E-mail, Aug. 2, 2007.

BIBLIOGRAPHY

Books and articles for further study on the subject of sensitive matter, as well as those used here as source materials.

GENERAL READING

Ball, P. *Made to Measure: New Materials for the 21st Century.* Princeton University Press, 1997.

de Gennes, P.-G. "Soft Matter." Nobel lecture in physics. *Review of Modern Physics,* 64 (1992): 645–648.

———, and J. Badoz. *Les objets fragiles.* Plon, 1994, and Pocket, 1999.

de Gennes, P.-G., and M. M. Veyssié. *Matière Molle.* CD-ROM. Microfolie's et Arte Éditions, 1998.

Guyon, E., and J.-P. Hulin. *Granites et fumées: Un peu d'ordre dans le mélange.* Éditions Odile Jacob, 1997.

———, and L. Petit. *Ce que disent les fluides.* Éditions Belin, 2005.

TECHNICAL SOURCES

Cabane, B., and S. Hénon. *Liquides, solutions, dispersions, émulsions, gels.* Éditions Belin, 2003.

Daoud, M., and C. Williams, eds. *La juste argile, collectif.* Les Éditions de Physique, 1995.

de Gennes, P. G. *Soft Interfaces.* Cambridge University Press, 1997.

———, F. Brochard-Wyart, and D. Quéré. *Gouttes, bulles, perles et ondes.* Éditions Belin, 2005. Reprinted with CD-ROM.

di Meglio, J.-M. "La matière molle." *Techniques de l'Ingénieur,* A1 195 (1994): 1–9.

Gerschel, A. *Liaisons intermoléculaires.* InterÉditions/CNRS Éditions, 1995.

Hamley, I. W. *Introduction to Soft Matter.* Wiley, 2000.

Jones, R. A. L. *Soft Condensed Matter.* Oxford University Press, 2002.

Kleman, M., and O. D. Lavrentovich. *Soft Matter Physics: An Introduction.* Springer, 2003.

Witten, T. A., with P. A. Pincus. *Structured Fluids.* Oxford University Press, 2004.

BY CHAPTER AND TOPIC

1. Peacemaking among Enemies . . . Easy When a Mediator Is Involved

Dickinson, E. *Les colloïdes alimentaires.* Masson, 1992.

Guyon, E., and J.-P. Hulin. "Le mélange non miscible," in Guyon and Hulin, *Granites et fumées: Un peu d'ordre dans le mélange.* Éditions Odile Jacob, 1997, pp. 193–204.

McGee, H. *On Food and Cooking: An Encyclopedia of Kitchen Science, History, and Culture.* London: Hodder and Stoughton, 2004.

Mertens, J. "Oil on Troubled Waters: Benjamin Franklin and the Honor of Dutch Seamen." *Physics Today,* 59 (2006): 36–41.

This, H. *Les secrets de la casserole.* Éditions Belin, 1993.

———. "Dérivés d'aïoli." *Pour la Science,* no. 243, 1998, 16.

———. "Ordres de grandeur." *Pour la Science,* no. 275, 2000, 13.

2. Dissolving Fat in Water: A Question of Organization

Hamley, I. W. "Amphiphiles," in *Introduction to Soft Matter.* Wiley, 2000, pp. 237–239.

Raisonnier, A. Digestion: Détoxification (digestion-detoxification). Medical school course, second term, first year, Faculté de Médecine

Pierre et Marie Curie, Université Paris-VI, 2003–2004. http://www
.chups.jussieu.fr/polys/biochimie/DGbioch/index.html.

3. Don't Mix, Associate!

Kelker, H. "Survey of the Early History of Liquid Crystals." *Molecular Crystals and Liquid Crystals*, 165 (1988): 1–43.

———, and P. M. Knoll. "Some Pictures of the History of Liquid Crystals." *Liquid Crystals*, 5 (1989): 19–43.

Mitov, M. *Les cristaux liquides. Que Sais-Je?* no. 1296, Presses Universitaires de France, 2000.

———. "Les cristaux liquides." *La Recherche*, 352 (2002): 48–51, and in *La physique en 18 mots-clés*, Dunod, 2009, 94–99.

Oswald, P., and P. Pieranski. *Les cristaux liquides*, vol. 1. Gordon and Breach, 2000, 36–39.

4. Rubber: A Story Nearly Cut Short

"The Charles Goodyear Story: The Strange Story of Rubber." Reprinted from *Reader's Digest*, January 1958. http://www.goodyear.com /corporate/history/history_story.html (accessed July 18, 2011).

Franks, F. *Polywater.* Cambridge: MIT Press, 1983.

Guinier, A. *La structure de la matière: Du ciel bleu à la matière plastique.* Hachette-CNRS, 1980, 238–241.

Institut de Formation du Caoutchouc. http://www.ifoca.com/spip/ (accessed July 18, 2011).

Institut de recherche en biologie végétale de l'Université de Montréal. http://www.irbv.umontreal.ca/ (accessed July 18, 2011).

Leperchec, P., and R. Malik. *Les molécules de la beauté, de l'hygiène et de la protection.* CNRS Éditions/Nathan, 1994.

Rousseau, D. L. "Case Studies in Pathological Science." *American Scientist*, 80 (1992): 54–63.

Staudinger, H. *From Organic Chemistry to Macromolecules.* Wiley, 1971.

5. The Firefighter's Jet Stream: Reach for the Sky

Amerchol Corporation. http://www.dow.com/PublishedLiterature/dh _005a/0901b8038005a373.pdf?filepath=amerchol/pdfs/noreg/324 -00009.pdf&fromPage=GetDoc (accessed October 20, 2011).

Collyer, A. A. "Demonstrations with Viscoelastic Liquids." *Physics Education*, 8 (1973): 111–116.

Ptasinski, P. K., F. T. M. Nieuwstadt, B. H. A. A. van den Brule, and M. A. Hulsen. "Experiments in Turbulent Pipe Flow with Polymer

Additives at Maximum Drag Reduction." *Flow, Turbulence, and Combustion,* 66 (2001): 159–182.

Truong, V.-T. *Drag Reduction Technologies.* Aeronautical and Maritime Research Laboratory, Fisherman's Bend, Australia, Defence Science and Technology Organisation, 2001.

6. The Glamorous Affair of Gas and Liquid

Ballet, P., and F. Graner. "Giant Soap Curtains for Public Presentations." *European Journal of Physics,* 27 (2006): 951–967.

de Gennes, P.-G. "Bulles, mousses . . . et autres objets fragiles." *Découverte,* no. 180, 1990, 43–60.

Graner, F. "La mousse." *La Recherche,* no. 345, 2001, 46–49.

Lagerge, S., ed. *Les mousses: Mouillage et démouillage.* EDP Sciences, 2002.

Liger-Belair, G. "D'où viennent les bulles de champagne?" *La Recherche,* no. 414, 2007, 79.

———, and G. Polidori. "Tempête dans un verre . . . de champagne!" *Pour la Science,* no. 362, 2007, 66–70.

Mousses et interfaces: La mousse déroule son film. Special issue, *Découverte,* no. 332, 2005, 22–60.

Peron, N., J. Meunier, A. Cagna, M. Valade, and R. Douillard. "Phase Separation in Molecular Layers of Macromolecules at the Champagne-Air Interface." *Journal of Microscopy,* 214 (2004): 89–98.

Vignes-Adler, M. "La mousse de champagne." *Bulletin de la Société Française de Physique,* no. 106, 1996, 27–29.

———, and F. Graner. "La vie éphémère des mousses." *Pour la Science,* no. 293, 2002, 48–55.

Wong, K., and D. Joubert. "Halte aux taches." *Pour la Science,* no. 266, 1999, 102–107.

7. Down with Foam!

Bergeron, V. "Antimousses et agents démoussants." *Techniques de l'Ingénieur,* J2 205 (2003): 1–5 and J2 206 (2003): 1–5.

Lagerge, S., ed. *Les mousses: Mouillage et démouillage.* EDP Sciences, 2002.

8. Breathing: An Unseen Triumph

"La corticothérapie anténatale." *Dossier du Centre National Hospitalier d'Information sur le Médicament,* 20 (1999): 11–12.

Courty, J.-M., and E. Kierlik. "A pleins poumons." *Pour la Science,* no. 289, 2001, 106–107.

Dietl, P., T. Haller, and S. Schneider. "Le surfactant pulmonaire." *Pour la Science,* no. 295, 2002, 84–91.

Fehrenbach, H. "Alveolar Epithelial Type II Cell: Defender of the Alveolus Revisited." *Respiratory Research,* 2 (2001): 33–46.

9. Familiarity and Distance: Colloids

Cabane, B., and S. Hénon. "Généralités sur les dispersions," in *Liquides, solutions, dispersions, émulsions, gels.* Éditions Belin, 2003, pp. 309–329.

di Meglio, J.-M. "Colloïdes et nanosciences." *Techniques de l'Ingénieur,* J2 130 (2007): 1–12.

Hamley, I. W. "Colloids," in *Introduction to Soft Matter.* Wiley, 2000, pp. 133–192.

Lafuma, F., and B. Cabane. "Gélification sous cisaillement." *Bulletin de la Société française de physique,* no. 60, 1993, 19–21.

10. Sensitive Cooking

Cabane, B., and S. Hénon. "Le droit au mélange." *Pour la Science,* no. 277, 2000, 106–107.

———. "Gels," in *Liquides, solutions, dispersions, émulsions, gels.* Éditions Belin, 2003, pp. 410–444.

Dickinson, E. *Les colloïdes alimentaires.* Masson, 1992.

———. "Colloid Science of Mixed Ingredients." *Soft Matter,* 2 (2006): 642–652.

Gagnaire, P. http://www.pierre-gagnaire.com/francais/cdthis.htm (accessed July 18, 2011).

Hamley, I. W. "Food Colloids," in *Introduction to Soft Matter.* Wiley, 2000, pp. 179–187.

Hertel, O. "La science passe à la casserole!" *Sciences et Avenir,* no. 707, 2006, 57–61.

Kalys Gastronomie. http://fr.gastronomie.kalys.com (accessed July 18, 2011).

Matricon, J. "La cuisine moléculaire," in *Graines de Sciences,* vol. 2. Le Pommier, 2000, 1–28.

Mezzenga, R., P. Schurtenberger, A. Burbidge, and M. Michel. "Understanding Foods as Soft Materials." *Nature Materials,* 4 (2005): 729–740.

This, H. "Les gels sont partout." *Pour la Science,* no. 327, 2005, 4.

————. "Modelling Dishes and Exploring Culinary 'Precisions': The Two Issues of Molecular Gastronomy." *British Journal of Nutrition*, 1352 (2005): S139-S146.

11. A Cell, Though Not a Prison

Auffray, C. "La cellule," in *Graines de Sciences*, vol. 1. Le Pommier, 1999, 171–193.

Goussard, J.-P. Système nerveux et muscles. http://caratome.free.fr/ (accessed July 18, 2011).

Grosbras, M.-H., and A. Adoutte. "La cellule." *La Recherche*, no. 288, 1996, 86–89.

Neurones 2011. http://www.neur-one.fr (accessed July 18, 2011).

12. Putting Drug Delivery on Controlled Release

Ball, P. "Clever Stuff: Smart Materials," in *Made to Measure: New Materials for the 21st Century*. Princeton University Press, 1997, pp. 134–142.

————. "Spare Parts: Biomedical Materials," in *Made to Measure: New Materials for the 21st Century*. Princeton University Press, 1997, pp. 236–241.

Brannon-Peppas, L. "Polymers in Controlled Drug Delivery." *Plastics and Biomaterials*, November 1997, 34.

Gerschel, A. *Liaisons intermoléculaires*, InterÉditions/CNRS Éditions, 1995, pp. 54–63.

Husseini, G. A., N. Y. Rapoport, D. A. Christensen, J. D. Pruitt, and W. G. Pitt. "Kinetics of Ultrasonic Release of Doxorubicin from Pluronic P105 Micelles." *Colloids and Surfaces B: Biointerfaces*, 24 (2002): 253–264.

Malmsten, M. "Soft Drug Delivery Systems." *Soft Matter*, 2 (2006): 760–769.

Uhrich, K. E., S. M. Cannizzaro, R. S. Langer, and K. M. Shakesheff. "Polymeric Systems for Controlled Drug Release." *Chemical Reviews*, 99 (1999): 3181–3198.

13. Perpetual Sensitivity: Granular Matter

Ball, P. *The Self-Made Tapestry: Pattern Formation in Nature*. Oxford University Press, 1999, 199–222.

Claudin, P. *La physique des tas de sable*. EDP Sciences, 1999.

————, and C. Reyraud. "Les tas de sable." *La Recherche*, no. 324, 1999, 86–89.

Constans, N. "Physique des grains: La science sur le tas." Special Issue, *Science et Vie,* no. 228, 2004, 142–151.

de Gennes, P.-G. "Granular Matter: A Tentative View." *Reviews of Modern Physics,* 71 (1999), S374–S382.

Duran, J. *Sables émouvants.* Éditions Belin, 2003.

———. "L'étonnante matière en grains." Special Issue, *Bulletin de l'Union des professeurs de physique et de chimie et de la Société française de physique,* no. 875, 2005, 67–73.

Guyon, E., J.-P. Hulin, and L. Petit. *Ce que disent les fluides,* Éditions Belin, 2005, pp. 138–151.

Guyon, E., and J.-P. Troadec. "Du sac de billes au tas de sable." Éditions Odile Jacob, 1994.

14. Liquefaction of the "Blood" of St. Januarius

Armelin, E., M. Marti, E. Rude, J. Labanda, J. Llorens, and C. Aleman. "A Simple Model to Describe the Thixotropic Behaviour of Paints." *Progress in Organic Coatings,* 57 (2006): 229–235.

Broch, H. "L'homme de l'art," in *Le Paranormal.* Seuil, Points Sciences, 1989, 107–111.

———. "La miraculeuse liquéfaction du sang de St Janvier," in *Au cœur de l'extraordinaire.* Éditions Book-e-Book, 2004, 310–314.

Courty, J.-M., and E. Kierlik. "Le Médium flamand." *Pour la Science,* no. 294, 2002, 106–107.

Coussot, P. "Rheophysics of Pastes: A Review of Microscopic Modelling Approaches." *Soft Matter,* 3 (2007): 528–540.

———, H. Van Damme, and C. Ancey. "Des solides coulants." *Pour la Science,* no. 273, 2000, 34–40.

Documentation Technique sur les Arts Plastiques et Avoisinants. http://www.dotapea.com/ (accessed July 18, 2011).

Garlaschelli, L. "Sangue prodigioso." *La chimica e l'industria,* no. 84, 2002, 67–70.

———, F. Ramaccini, and S. Della Sala. "Working Bloody Miracles." *Nature,* 353 (1991): 507.

———. "A 'Miracle' Diagnosis." *Chemistry in Britain,* 30 (1994): 123–125.

Jorio, T. P. *La Chapelle royale du trésor de saint Gennaro,* trans. C. Putzier. Museo San Gennaro, Naples, 2002.

Khaldoun, A., E. Eiser, G. H. Wegdam, and D. Bonn. "Liquefaction of Quicksand under Stress." *Nature,* 437 (2005): 635.

Larcher, H. *Le sang peut-il vaincre la mort?* Gallimard, 1957.

Moller, P. C. F., J. Mewis, and D. Bonn. "Yield Stress and Thixotropy: On the Difficulty of Measuring Yield Stresses in Practice." *Soft Matter,* 2 (2006): 274–283.

Oswald, P. *Rhéophysique ou comment coule la matière,* Éditions Belin, 2005, 24–31.

Strazzullo, F. *St. Gennaro's Treasure,* trans. I. Sisero. Chapel of the Treasury, Naples, 1992.

Bonus Tracks

Jensen, P. *Entrer en matière.* Seuil, Points Sciences, 2004.

Lévy-Leblond, J.-M. *Aux contraires: L'exercice de la pensée et la pratique de la science.* Gallimard, 1996.

Quéré, Y. "Les matériaux," in *Graines de Sciences,* vol. 1. Le Pommier, 1999, 75–114.

ACKNOWLEDGMENTS

For giving me encouragement and for providing answers to my questions, I sincerely thank Ferran Adrià, Daniel Bonn, Georges Bossis, Yves Bouligand, Henri Broch, Jean-Claude Carrière, Philippe Chovet, Philippe Claudin, Xavier d'Arodes de Peyriague, Pierre-Gilles de Gennes, Monsignore Vincenzo De Gregorio, Elodie Escaig, Jacques Friedel, Jean-Marc Lévy-Leblond, Guy Molénat, Donald M. Small, Hervé This, Madeleine Veyssié, and Nicolas Witkowski.

CREDITS

Pages 6, 19, 25, 26, 30, 38, 54, 56, 62, 72, 98, 101, 107, 109, 110, 123, 127: © M. Mitov. Page xv: Coll. part. M. Mitov. Page 14: From ch. 8.16 of the course "Digestion—Détoxification," taught by Prof. A. Raisonnier, Pierre and Marie Curie Faculty of Medicine, Pitié-Salpêtrière. Page 22: With the kind permission of J. Friedel. Page 24: © S. Coetsier. Page 29: From "PDLC (Polymer Dispersed Liquid Crystal) Nemâticos," with the kind permission of P. J. O. Sebastião, University of Lisbon. Page 80: From figure 19, "Microcalorimétrie à balayage DSC," P. Relkin, *Techniques de l'Ingénieur,* P1270, pp. 1–16, 2006. Pages 94, 96: From "General and Human Biology;" © The McGraw-Hill Companies, Inc. Page 103: © CNRS Photothèque/Sagascience/F. Caillaud. Pages 104, 106: From A. Gerschel, *Liaisons intermoléculaires,* InterÉditions/CNRS Éditions, pp. 60, 62, 1995. Page 135:

INDEX OF SENSITIVE MATERIALS

INDEX OF PROPER NAMES